KV-371-253

Ethical eye

French version:

Regard éthique – La recherche biomédicale

ISBN 92-871-5461-9

Cover design: Graphic Design Workshop, Council of Europe
Cover illustration: photo from Getty images
Layout: Pre-Press Unit, Council of Europe
© Council of Europe

Edited by Council of Europe Publishing
http://book.coe.int
Council of Europe Publishing
F-67075 Strasbourg Cedex

ISBN 92-871-5462-7
© Council of Europe, October 2004
Printed in Germany by Koelblin-Fortuna-Druck

Contents

EUROPE AND BIOMEDICAL RESEARCH

European law and biomedical research

List of contributors

Claude Huriet

Claude Huriet is a professor at the Faculty of Medicine in Nancy, an honorary member of the senate, and President of the Institut Curie. He was formerly a member of the French National Consultative Committee on Ethics.

Stéphane Bauzon

Stéphane Bauzon has a doctorate in Law from the University of Paris-II and a diploma from the Institute for Political Studies in Strasbourg. He is a Lecturer in the Philosophy of Law at the University of Rome "Tor Vergata" and a scientific expert on the Italian Bioethics Committee.

Michel Coleman

Since 1995 Michel Coleman has been a Professor of Epidemiology and Vital Statistics at the London School of Hygiene and Tropical Medicine, and Deputy Chief Medical Statistician at the Office for National Statistics. He has worked at the WHO International Agency for Research on Cancer in Lyons (1987-91) and was Medical Director of the Thames Cancer Registry in London (1991-95).

Tom Gallacher

Tom Gallacher is a Director of Medical Policy and Standards for GlaxoSmithKline. He is a Board Member of the European Forum for Good Clinical Practice and a member of the Ethics Committee of the International Federation of Pharmaceutical Manufacturers Association.

Eugenijus Gefenas

Dr Eugenijus is an associate professor and director of the Department of Medical History and Ethics at the Medical Faculty of Vilnius University. He is also Chairman of the Lithuanian Bioethics Committee, and a senior research fellow at the Institute of Culture, Philosophy and Arts. Dr Gefenas is a member of the Council of Europe Bioethics Committee (CDBI) as well as a member of the International Bioethics Committee (IBC) of Unesco.

Andrzej Górski

Professor of Immunology and Medicine, Andrzej Górski is the Director of the Hirszfeld Institute of Immunology and Experimental Therapy, Polish Academy of Sciences and a professor at the Medical University, Warsaw. He has published over a hundred papers in international scientific journals and has won numerous awards for his work, including the Meller Award "for excellence in cancer research".

Vivienne Harpwood

Vivienne Harpwood has a chair in Law at Cardiff University, and is also a barrister with experience in practice. She is Director of the Centre for Medico-Legal Studies, based at Cardiff Law School. Her main research interests are medical law and the law of tort, and she has published extensively in these areas. In 1987 she established the LLM (Legal Aspects of Medical Practice) degree course at Cardiff University, which attracts students, healthcare professionals and lawyers from all over the world.

Outi Leena L. Hovatta

Outi Leena Hovatta is a graduate of the University of Helsinki. A specialist in Obstetrics and Gynaecology, she was also the founder of the Infertility Clinic of the Family Federation of Finland, the largest *in vitro* fertilisation clinic in Finland for the time being. From 1995 to 1998 she was a visiting scientist and visiting professor at Imperial College School of Medicine, London, and from September 1998 has been Professor of Obstetrics and Gynaecology in the Karolinska Institutet, Stockholm.

Herman Nys

Herman Nys obtained a master's degree and a doctorate in law at K.U. Leuven. He specialised in medical law at European universities (Nijmegen; London) and he now teaches medical law in the Medical and Law School of the K.U. Leuven and has been a guest professor at the Université Catholique de Louvain. He is also professor in International Health Law at the University of Maastricht in the Netherlands.

Povl Riis

Professor Riis was a Professor of Medicine at the University of Copenhagen from 1974 until 1996. During that same period he was also Head of the Medical Department of Herlev University Hospital. From 1972 until 1974 he was Chairman of the Danish Medical Research Council and from 1979 until 1998 he was a member of the editorial board for the *Journal of the American Medical Association.*

Maxime Seligmann

Dr Maxime Seligmann is a Professor Emeritus of Immunology at the University of Paris-VII. For twenty years he has been Head of the Inserm Research Unit on Immunochemistry and Immunopathology, and Head of the Clinical Department of Immunopathology and Hematology at Hôpital St-Louis Paris. He is a member of Embo, and currently a member of the French National Ethics Committee.

Jan Helge Solbakk

Professor Jan Helge Solbakk is a physician and a theologian and also holds a PhD in ancient philosophy. Currently he is Professor and Director of the Section for Medical Ethics, Faculty of Medicine, University of Oslo. He is also Adjunct Professor of Medical Ethics and Philosophy of Medicine at the Centre for International Health, Faculty of Medicine, University of Bergen.

Sreeharan

Dr Sreeharan is currently Senior Vice-President and Post Graduate Medical Director of the GlaxoSmithKline Academy of Pharmaceutical Medicine. He was previously Senior Vice-President and European Medical Director for a number of years. Prior to joining the pharmaceutical industry, he held a number of teaching and professorial positions, including the Foundation Chair in Medicine at the University of Jaffna, Sri Lanka.

Jochen Taupitz

Dr Jochen Taupitz is Professor at the University of Mannheim and Managing Director of the Institute for Medical Law, Public Health Law and Bioethics of the Universities of Heidelberg and Mannheim.

Pēteris Zilgalvis

Pēteris Zilgalvis, J.D. is Deputy Head of the Council of Europe's Bioethics Department and Secretary of the Working Parties on Biomedical Research, on Biotechnology and on Research on Human Biological Materials. He is the author of twenty publications in books and journals on bioethics and environmental law, as well as of numerous magazine and newspaper articles on bioethics, the environment and economic reform. He is also a member of the editorial board of the *Baltic Yearbook of International Law*.

Preface

The fifth book in the Ethical eye series focuses on biomedical research, a field involving a number of fundamental rights. On the one hand, there is the right to freedom of research, established in the Convention on Human Rights and Biomedicine, as well as in other international instruments and a number of national constitutions. The freedom of science has helped to produce and promises to produce great advances in knowledge, which translate into practical applications in the field of healthcare, for instance.

On the other hand, there are the rights of individuals: protection of their dignity, identity and integrity. The Council of Europe has always emphasised the primacy of the interests and welfare of the human being over the sole interest of science or society.

Thus, biomedical research is a sphere in which it is necessary to find an equilibrium between these different rights. There cannot be unrestrained freedom of research without regard to other fundamental rights, just as one cannot ignore the interests of society or science in our European culture based on solidarity. Researchers, medical professionals, ethicists, philosophers, lawyers and policy makers must undertake a multidisciplinary dialogue in order to find this equilibrium. This dialogue takes place in research settings, hearings, ethics committees, parliaments and international fora, such as the Council of Europe.

Science is, by its very nature, international and is not isolated within the borders of any single state. Research projects are often undertaken in a number of countries at the same time and researchers regularly go abroad to undertake research. This very mobility can give rise to ethical concerns. In any case, modern communications speed news of discoveries to all corners of the globe. It is therefore important that ethical rules are identified, endorsed and applied by all the scientific community and subject to public debate. International instruments and fora contribute to reaching these goals.

The Council of Europe has adopted an Additional Protocol to the Convention on Human Rights and Biomedicine on Biomedical

Research. This protocol addresses both basic and clinical research. Based on the Convention, it develops rules for the protection of individuals, especially by giving guidance to the research ethics committees. It also provides additional protection for those who are unable to consent and those who are in a vulnerable position such as prisoners. As the Parliamentary Assembly highlighted, the protocol also contains rules for transnational research, to ensure that its principles also apply to research projects undertaken by European firms or researchers that will be carried out abroad.

Elaborating ethical and legal norms on research is just a first step. The next one is to have those norms applied. For this purpose, it is important that researchers and other stakeholders take part in a constructive debate. The creation of a European Bioethics Forum in Strasbourg would no doubt contribute to obtaining a high quality debate. This project is being developed by the European Democracy Forum and supported by the City of Strasbourg and the Council of Europe. It would create an environment for dialogue open to the largest possible number of actors and where ethical issues would be approached in a multidisciplinary and pluralist way.

This edition in the Ethical eye series is another part of this dialogue, bringing together just such a multidisciplinary group of authors from different countries in Europe to discuss different approaches, issues, achievements and problems in this field of science and human rights.

The importance of biomedical research is clear, and it is to be hoped that this book will contribute to both the support for science and specifically for biomedical research in Europe and to the protection of those persons who participate in research projects.

Introduction

by Claude Huriet

The background

The Act of 20 December 1988 on the Protection of Human Subjects of Biomedical Research, which was – if it need be pointed out – the first piece of legislation on clinical research, had scarcely been passed by the French Parliament when complaints rang out in many quarters about lawmakers interfering in areas they should keep out of.

The same prophets of doom predicted that the act would be a severe blow to biomedical research in France, with the promoters of clinical trials moving their activities to other countries to escape the constraints of the new legislation.

While the act was largely based on ethical principles that had already been proclaimed, it was clear that these established principles – set out forcefully in Article 1 of the 1947 Nuremberg Declaration, which described voluntary consent as being absolutely essential, and in subsequent international declarations, before being taken up in one of the first opinions issued by the French National Advisory Committee on Ethics in 1984 – had not really been taken account of, unless, of course, people had immediately realised that the discrepancy was the result of the difference between ethical recommendations, which do not involve obligations, and legislative provisions, contravention of which is subject to penalties.

Fortunately, as reasoned optimism had suggested would happen, neighbouring countries soon recognised the risk of questionable or unacceptable biomedical research practices, of the kind banned under the French legislation, developing within their own borders. The Council of Europe responded very speedily with Recommendation R(90)3 concerning medical research on human beings, the fundamental principles of which were clarified in the Convention on Human Rights and Biomedicine (see chapter by Zilgalvis). I saw this for myself at a meeting of the Debra[1] co-operation programme in Vilnius.

1.
The Debra co-operation project, set up in 1997, fosters the development of independent and multidisciplinary ethics committees for review of biomedical research in central and eastern European countries.

The speed of this response, even though it did not involve binding texts and even though it took almost ten years for the joint declarations to be finalised, clearly sent out a strong signal that discouraged the introduction of unethical practices in a particularly sensitive area.

The fact that Council of Europe Publishing is now bringing out a publication on human rights and biomedical research fits in logically with the work first started fifteen years ago.

The book highlights the ethical challenges of research and provides an overview of biomedical research in Europe, before setting out the legal instruments of the Council of Europe, whose contribution in this area has been considerable, as well as those of other European institutions in the biomedical research field.

Three key points emerge from the various chapters written by different authors, of different nationalities:

• medical research on human beings is based on shared values;

• efforts need to be made to agree shared definitions or give shared concepts identical meanings;

• although the publication is made up of European contributions, many of the authors place their views in the broader context of the globalisation of research.

Shared values

Any discussion of research on human beings draws on the Nuremberg Code, which formed part of the judgment by the international tribunal that tried and sentenced Nazi criminals for the human experiments carried out in the concentration camps.

The following preconditions apply to biomedical research:

• it is absolutely essential for the persons concerned to give their voluntary consent after being properly informed about the research;

• the experiments must yield fruitful results for the good of society;

• it is essential for certain scientific prerequisites to be met: for the scientists involved to be properly qualified, for the risk–benefit balance to be assessed and for all unnecessary physical and mental suffering and injury to be avoided.

These fundamental values are set out in human rights texts (see chapter by Solbakk) such as the United Nations International Covenant on Civil and Political Rights (Article 7).

However, having set forth these principles, it has to be admitted that it is not always easy to put them into practice. Obtaining consent for participation in clinical trials is impossible in certain circumstances; for example, in certain serious situations such as emergencies or even in cases of Alzheimer's disease. Even preserving medical confidentiality, a widely held deontological obligation can sometimes, in the field of research, be at odds with methodological rigour and public health issues. A clear case in point is the contradictory situation in the case of cancer registers (see chapter by Harpwood and Coleman).

Thus it has to be recognised that despite repeated assertions that the patient's interests are of paramount importance above and beyond those of science and society, sometimes the benefit to the greatest number wins the day. In such situations ethical considerations are all the more necessary and nuanced, and the term "ethical tensions" all the more apposite.

Apart from the possible pitfalls of the inappropriate translation of abstract concepts, the mere use of the same terms does not mean that the thinking or the content concerned are identical. This is illustrated by two examples: ethics and ethics committees.

Shared definitions

Focusing an ethical eye on biomedical research is inconceivable without first defining "ethics". This is all the more essential since there is a whole range of definitions, a fact which is a source of confusion and, in some cases, of contradictions.

Confusion between the deontology and ethics of research may be surprising. Even so, before considering individuality ethics or collectivity ethics (see chapter by Riis) or making a distinction between the ethics of clinical research and the ethics of clinical care (see chapter by Górski), I would suggest that we need agreement on the following characteristics of ethics in the biomedical field.

It is an evolving process that does not claim to be universal. Being pluralist in essence, it aims to achieve an acceptable balance at both individual and collective levels between the protection of individual dignity and freedom and the expectations of society.

The non-universal nature of ethics stems from the fact that, although ethical perspectives are based on universal values of a spiritual, religious and/or metaphysical character, they also take account of historical and cultural references, and even the economic environment.

While the principle of prior consent to any biomedical research is generally accepted, the conditions for obtaining it, which ethics is supposed to govern, are variable. This was brought home to me by a visit to Mali where I had gone to talk about the ethics of research. "Consent" there is given by the village headman, who receives the compensation granted for the constraints the research imposes on the village population. This practice is regarded as "unethical" in our part of the world … and yet!

In central and eastern Europe (see chapter by Gefenas), participation in clinical trials can enable the individuals concerned to receive treatment to which they would not otherwise have access.

Another way of obtaining consent is illustrated by the national genome project in Estonia, with reference to unconditional overall consent rightly being regarded as unethical.

The questioning that characterises the ethics process and the non-universality of ethics should be seen alongside the very object and purpose of the process. After all, ethics bodies draw up recommendations – not rules or laws. They issue opinions but do not produce whole sets of rules that apply to everyone and involve penalties if they are broken. From this point of view, I fear that the references sometimes made to "non-binding rules" may be a source of confusion.

In their diversity, ethics committees, whether public entities or private-law bodies (see chapter by Taupitz), also reflect the non-universal nature of ethics. Apart from the fact that, although they vary in composition, there are no countries where they are democratically elected (for which they are sometimes criticised), some are purely advisory, while others do have decision-making powers (see chapter by Gallacher and Sreeharan). Reference may be made here to the study conducted by the Lithuanian bioethics committee (see chapter by Gefenas) on obtaining informed consent.

The discussions about the use of placebos* clearly demonstrate the wide range of opinions and arguments based on an ethical approach and how the public can actually be involved, provided a little effort is made.

One cannot but agree with the interpretation that giving up placebo-controlled clinical trials* is just as unethical as using them inappropriately, which Jean Bernard summed up in the phrase "everything that is unscientific is unethical".

In contrast, we are bound to be surprised by the National Placebo Initiative organised in Canada (see chapter by Górski), which, to my knowledge, is the first attempt to involve the public in ethical choices.

The wide range of situations, the great differences in cultural and economic environments, the large number of texts of varying legal force and the difficulties that can arise in imposing penalties on researchers who violate human rights in the name of medical research (inappropriate research, inappropriate data acquisition or usage, inappropriate incentives, and so on; see Gallacher and Sreeharan), have led to calls in some quarters for the establishment of an international tribunal based on an international forum that could help harmonise practices and prompt measures for overcoming the terrible inequality that affects poorer nations (see chapter by Solbakk).

Globalisation of research

Although this is a European publication, the globalisation of research is mentioned in various chapters.

Placebo:
an inactive pill, liquid, or powder that has no treatment value. In clinical trials, experimental treatments are often compared with placebos to assess the treatment's effectiveness.

Placebo-controlled clinical trials:
a method of investigation of drugs in which an inactive substance (the placebo) is given to one group of participants, while the drug being tested is given to another group. The results obtained in the two groups are then compared to see if the investigational treatment is more effective in treating the condition.

Reference to the "globalisation of research" actually involves two separate trends (see chapter by Riis): increasingly frequent, international multi-centre studies to recruit more easily and in short periods cohorts for the clinical study of rare diseases, among other examples, and co-operation between developed and developing countries.

For people in developing countries, there is a significant risk of developed countries "exporting ethical problems and importing research findings". The terrible global inequality (see chapter by Solbakk) is illustrated by the following statistics. In 1996, 90% of the US$56 billion spent on medical research concerned the needs of the richest 10% of the population and the illnesses that affect them.

Apart from respect for the particular cultural, religious, economic and sociological features of developing countries, special attention must be paid to such questions as to the way informed consent is obtained among illiterate populations, the determination of control groups, the reference treatments or placebos, the diseases studied (malaria, tuberculosis, parasitic diseases and Aids, among others) and the continuation of the treatments after the completion of the trials.

The complexity of these situations – and the overriding need to reverse or prevent the implementation of ethically unacceptable practices – led the Council of Europe to supplement the Convention on Human Rights and Biomedicine, adopted on 19 November 1996, with an additional protocol on biomedical research. Research conducted outside the member states is covered in Article 29. Under its provisions, researchers may not act "as they see fit" in countries where standards for the protection of human beings in the research field are low or non-existent (see chapter by Zilgalvis).

A phrase that preface writers and literary critics are wont to use – and overuse – is that a work is very timely. That certainly is true of this book by Council of Europe Publishing, made up of contributions from recognised experts.

In spite of its noble aims, biomedical research, which is a source of progress in understanding the human being and improving individual and public health, must not be allowed to develop without any account being taken of ethical considerations or legal rules, as if "the end justified the means".

Such noble aims must also not be pursued for the sole benefit of only a part of the world's population. Biomedical research must involve a socially committed approach based on respect for human rights, to which the Council of Europe is contributing through this publication, as that is its *raison d'être*.

History and definitions

by Povl Riis

The history of human rights and ethics relating to research on human beings is fairly long, but limited in scope, if we take into account only research based on evidence and not prejudices and religious beliefs.

Well-known figures such as Ambroise Paré (1510-90), who ended the treatment of gunshot wounds with boiling oil, replacing it with the use of ointment, James Lind (1716-94), who described the effect of lemon juice in the prevention of scurvy, Johannes Fibiger (1867-1928), who made a systematic comparison of serum treatment of diphtheria and treatment without serum, and Austin Bradford Hill (1898-1991), who was in charge of the controlled trial demonstrating the effect of streptomycin on pulmonary tuberculosis, to mention just a few forerunners of modern scientific methodology, were primarily inspired by the need to discover the truth about the potential effects of known or new biomedical practices, an additional motivation being to help the soldiers, sailors, children and tuberculosis patients concerned.

However, despite any optimism that may prevail concerning human and societal developments, it is the transgression of ethical boundaries that has been the key incentive for devising and applying ethical concepts and standards relating to biomedical research, codes of good practice in this field and control systems based on the establishment of research ethics committees.

In other words, the development of high ethical standards has historically been less a progression phenomenon, bound up with greater prosperity and general political progress, but more a transgression phenomenon, as a reaction to severe disregard for fundamental human rights. As a consequence of the shift from transgression to progression, the primary focus today is the key figure of biomedical research ethics, the patient or the healthy volunteer, whose safety and right to respect and autonomy must be guaranteed.

The need for definitions

Ethics as a term in itself – that is, unrelated to research – has a much longer history than research ethics. "Biomedical research ethics" is accordingly a new expression, which has been part of scientific terminology for only the last four to five decades. As a recent addition to scientific language, this expression has caused much conceptual confusion and a correspondingly strong need to define the term or, at least for its users, to clarify the meaning they attach to it. In order to reduce the scope for confusion, the key terms of research ethics are defined below.

To leave etymology behind and focus instead on semantics, ethics – in the sense I accept for it – can be defined as follows:

> An overall term for the immaterial values, norms and attitudes prevalent in a country or culture, which underlie that country's or culture's concept of humankind and the laws and codes based thereon and shape citizens' personal existence and relations with each other and with the legal and private institutions of society. From a global perspective, ethics also includes responsibility for the ecological balance of planet Earth, its soil, water and air and its biological diversity.

After the Second World War and the large-scale serious violations of human rights, even in research, perpetrated in the concentration camps, prisons and ghettos, the main focus of ethical codes and supervisory measures was the human being as an individual, with strong emphasis on rights. I call this part of ethics "individuality ethics". It is still a fundamental, inescapable part of ethics.

However, it often overshadows our important responsibility for our fellow human beings, sometimes known as distributional ethics, but which I call "collectivity ethics", to use a less technical term. Even if the starting point for codes and for research ethics committees is the safety of and respect for individual trial participants, they also need to consider the societal aspects of solidarity and altruism, where these are based on genuine informed consent. There could be no scientific progress if the population of Europe did not have a latent sense of collectivity ethics, not only among healthy volunteers but also among participants in randomised drug trials,* in phase-one and phase-two drug trials, in epidemiological projects, in genetic family studies and in many other research areas.

The ethics of research, as an overall term for all ethical aspects of biomedical science, has until recently been taken to mean research ethics concerned with the safety of and respect for research subjects. Research ethics is, however, a *toto pro parte* term (an overall term applied to a part), which from a linguistic standpoint excludes another important branch of the ethics of research: researchers' ethics, that is, good-practice standards concerning the reliability of harvested variables, interpretation of data, trustworthiness of publications and respect for other scientists' intellectual property.

This second part of the ethics of research will not be discussed at length here, although ethical good conduct and reliability are conditions for obtaining genuine informed consent. In Europe, experience has shown that codes and supervisory agencies with fraud-prevention aims are best established independently of those dealing with research ethics, yet with some sort of co-ordinating link between the two, and this argues against discussing that branch of ethics here.

Biomedicine comprises all the disciplines related to the health services and their educational institutions. As we will see later, the ethics of biomedicine also covers projects carried out by non-health professionals, if they obtain access to patients and make their diagnoses via the health system, its case records and other data.

The term "intervention" is used here to mean any planned measurable influence on a person's mind or body. It comprises interviews, cognitive tests, diagnostic tests, surgery, drug therapy, preventive arrangements, information about serious life conditions and events, and so on. Use of the term "intervention" makes it possible to avoid the usual excessive reliance on the narrower term "therapy".

Research subjects can be healthy volunteers, participating patients or, in some countries, fertilised human eggs, foetuses and even deceased persons.

Randomised drug trials: these consist of a study in which participants are randomly (i.e., by chance) assigned to one of two or more treatment arms of a clinical trial. Occasionally placebos are used.

Pathogenesis:
the manner of development of a disease.

The term "human rights" in a contemporary context, as used here, refers to the international (including European) declarations, directives, conventions and similar codes.

Ethical codes, apart from those mentioned above, include a large number of United Nations declarations and professional bodies' declarations, guidelines and the like, in addition to national law on research ethics. There are now a very large number of codes worldwide, which can sometimes be confusing for biomedical scientists, as some of these codes even contradict each other. The final section of this chapter suggests a hierarchical order for these many codes.

Ethical committees, in the accepted sense used here, mean ethical research committees dealing with biomedical research involving human beings, excluding such committees' advisory functions not related to research, for instance giving advice on therapeutic dilemmas.

The scope of biomedical research

During the last three to four decades the scope of biomedical research has widened considerably, with a proportional broadening of the ways in which research ethics must be applied. Even if the starting point for any research ethics approach is the individual patient in a clinical setting, the scope is very extensive.

It ranges from molecular biological studies of pathogenesis,* to clinical and genetic studies and to epidemiological projects, covering individual nations (regions) or more than one country. Research ethics form part of the ethical aspects of a project not only where live patients participate in a project, but also where the source of variables is biological material from former patients. The same may apply where the intention is to use existing confidential data or new personal information of a sensitive nature, to be obtained via questionnaires.

The globalisation of biomedical research

During the last two to three decades, not only has the scope of biomedical research been extended as a result of the ever-broader spectrum of methods and methodologies, but the universe of scientists of many types – disciplinary globalisation – and the number of nations collaborating in research and implementing the results – geographical globalisation – have also grown considerably. This development has brought about a corresponding increase in demand for ethical codes and supervisory committees, whereas even significant post-Second World War codes remained centred on the medical practitioner, a clinician, and his or her patient.

Today, many disciplines are grouped together in a growing number of multidisciplinary teams: pharmacists, dentists, nurses, biochemists, biological geneticists, midwives, social workers, sociologists, psychologists and others, in addition to physicians. If all such disciplines devise their own professional codes (and some have shown signs of doing so), valuable multidisciplinary projects will be jeopardised by delays and sheer bureaucracy. Conversely, multidisciplinary research will gain from the emergence of national and European common codes, and, at the same time, the fundamental human rights behind each discipline's ethics will remain unchanged.

Geographical globalisation covers both international multi-centre studies and developed countries' cooperation with developing countries. International multi-centre studies, implemented by equal partners, are necessary to identify quite small but significant therapeutic benefits in the treatment of common diseases, such as cardiovascular diseases, because the groups compared have to be very large. Similar international co-operation is also necessary for clinical study of rare genetic diseases.

The second type of geographical globalisation, co-operation between developing and developed countries, raises special research ethics issues, which existing codes until recently failed to address. Projects of this kind will in future have to afford the populations of developing countries better protection against exploitation – in the form of the "export of ethical problems and import of results" – in connection with interventions important to the developed countries. There is an obvious need to involve anthropological advisers in the planning phase, in order to respect and take into account national religious, cultural and political norms (while naturally ensuring full respect for fundamental human rights).

Other major research ethics problems, not answered by the "Just do as we do" attitude of affluent research partners, include:

- obtaining genuine informed consent in illiterate groups and/or those with a gender imbalance;

- the interventions offered to control groups, whether a placebo or an alternative, hitherto-used intervention, in order to comply with the ideal, but unrealistic, rule that control groups must receive the best intervention known worldwide, irrespective of cost and local accessibility;

- what to do when the research is finished, as a fixed item on the agenda of planning meetings;

- competence building and co-authorship for participating professionals from the developing country, instead of "the flying scientist" from the developed country gathering the necessary results and then leaving on the first outward-bound plane, making away with the loot for his or her own benefit;

- and, probably the most important ethical rule, to ensure that both developing and developed countries' ethical committees have approved the project with a mutual right of veto.[1]

Following the preparatory work done by the Nuffield Foundation,[2] it is to be hoped that a European protocol governing this area of research ethics will be drawn up in future.

Research ethics committees

Some countries, including the new European democracies, have still not established research ethics committees with structures and functions that facilitate inter-European and more wide-ranging international cooperation. The positive side of this is that they can make the most of their thirty to forty years' experience with existing committee systems, but the downside is that these existing committees constitute an obstacle to the implementation of valuable inter-European projects.

1.
The host country can always veto the research in its jurisdiction, and, while the "exporting" country can veto participation by its own researchers or companies, it does not exercise any other form of control over research in the host country.

2.
Nuffield Foundation: founded by William Morris (Lord Nuffield) in 1943; for more information see: http://www.nuffield foundation.org/

Countries wishing to establish a committee system or to revise an existing system have to answer a number of key questions before deciding on the structure to be adopted.

Do they want a semi-official system or a law-based system? If policy makers consider a law-based system acceptable, it will probably facilitate international co-operation. Moreover, this implicitly answers the next question: Should there be a nation-wide system of regional committees or a network of mutually independent institutional committees? The latter solution may appear to be the easiest but will, at the same time, also prove the least satisfactory, according to the well-known adage that what is next to best is the best's worst enemy. Will the system be single- or two-tiered (either a number of committees with authority to take final decisions or a system of regional committees with a central committee whose role consists in hearing appeals, co-ordinating matters and advising the government, parliament, ministries and so on)?

Given the importance of geographical globalisation, the next question is: Will our national system be geared to deal not only with national projects but also with co-operative projects with other developing countries?

Probably the most challenging question is: Will lay persons sit on the committees and, if so, how many? And how will such members be elected? One possibility is for lay members to be elected/nominated through a representative democratic process, and for scientific members to be chosen under a similar procedure within scientific fora (academies, scientific societies, research councils).

How will the costs be met? In a two-tiered system, will local committees be financed via the regional authorities, and the central committee via the relevant ministries (health, research and/or education)? And will members be salaried? Will applicants have to pay a fee?

Research ethics committees should in any case be established in accordance with the requirements and conditions set forth in the Council of Europe's Convention on Human Rights and Biomedicine and its explanatory report, the protocols thereto and their explanatory reports, and the EU directives.

Randomisation:
a method based on chance by which study participants are assigned to a treatment group. Randomisation minimises the differences among groups by equally distributing people with particular characteristics among all the trial arms. The researchers do not know which treatment is better. From what is known at the time, any one of the treatments chosen could be of benefit to the participant.

Blinding:
a randomised trial is "blind" if the participant is not told which arm of the trial he is on. A clinical trial is "blind" if participants are unaware of whether they are in the experimental or control arm of the study.

Research ethics and its anchoring in society

One important aspect of research ethics committees' work is the contribution they make to public understanding of the need for biomedical research and of the fact that participation by patients and healthy volunteers is inevitable, if the population at large and specific patient groups are to benefit.

Terms such as evidence-based medicine, trials, randomisation,* blinding,* placebo, bias and many others are foreign to many – even well-educated – citizens and, because the media usually fail to bridge this language gap, it is necessary for chairs and members of committees to reach out to the public via the spoken or printed word. It takes time to counter the popular myth, sometimes sustained by the media, that doctors do research only because it will advance their careers (which is part of the reason, but certainly not the full explanation). This can be achieved by explaining how personal impressions or unsystematic experience have sometimes resulted in prolonged use of ineffective or even dangerous interventions. In other words, research based on good scientific methodology, evaluated by independent research ethics committees, must be considered a public benefit. From the globalisation standpoint, it allows individual countries to contribute to human well-being, while at the same time reaping considerable benefits from other countries' research. The examples of this global exchange are legion, but its benefits have yet to be demonstrated to the public.

The diversity of codes and Europe's role

Since the Second World War the number of international and national codes of research ethics has grown tremendously, reflecting a virtual explosion of interest in and commitment to ethics, and the sub-group of research ethics. Many of these codes are intended to raise scientists' and scientific societies' awareness of the issues at stake and to provide them with guidelines. On the one hand, this is a positive phenomenon but, on the other hand, it sends a confusing cacophony of signals to planning and project researchers. Even in the case of monocentre, national projects it can be confusing if national, regional or international codes fail to adopt the same viewpoints or to set the same ethical standards, but this more frequently poses a dilemma where biomedical research is based on inter-European or more wide-ranging international multicentre studies, not to mention interdisciplinary projects, which are often combined with international studies.

It is accordingly appropriate and necessary for scientists to be familiar with the multitude of codes, but also to be able to rank them by order of importance and binding authority. A brief hierarchical ranking, based on a number of principles, is set out below as guidance for European biomedical scientists. The aim is also to encourage European scientific societies and policy agencies to extend European initiatives towards the adoption of additional codes, in the form of conventions and directives, so that Europe speaks with a single voice in global debate and decision-making on research ethics.

Historically, the World Medical Association's Declaration of Helsinki (see Appendix III) strongly influenced post-Second World War interest in and respect for biomedical research ethics. The latest (fifth) version unfortunately fails to take account of the huge expansion of the research sector's scope, its changing methodologies and its disciplinary and geographical globalisation (see the references at the end of this chapter). This applies in particular to the paragraphs dealing with what should happen when research is over and the considerations to be borne in mind when deciding the interventions for a randomised control group. Making it an ethical standard that in a controlled trial the best intervention known must be offered to the control group, irrespective of costs and national accessibility – and imposing a similar idealistic, but sometimes unrealistic, condition for treating all participants once the research is over – can have the paradoxical effect, despite good intentions, of blocking much needed co-operative biomedical research in the poorest developing countries. Despite these shortcomings, the declaration is still an inspiring document, although it has now been overtaken by, for instance, the European Convention on Human Rights and Biomedicine and its protocols and the EU directives.

On the international and global scene, the UN codes are also important for European researchers. Unesco's Universal Declaration on the Human Genome and Human Rights deals with genetic research, as the title suggests, and cannot but have a strong advisory influence on European scientists (see the bibliography at the end of this chapter).

The same can be said of WHO's work on research ethics, especially that done by the Council of International Organizations of Medical Sciences (CIOMS), which recently published its Guidelines for Biomedical Research involving Human Subjects. An example of a special code is UNAIDS' "Ethical considerations in HIV preventive vaccine research".

The United States, through the National Institute for Health, has issued *Guidelines for the Conduct of Research Involving Human Subjects* at the NIH, which are applicable to both US research and other countries' research sponsored by the USA.

As mentioned earlier, the most important code outside the United States, dealing with biomedical research in developing countries supported or performed by developed countries, is the book issued by the Nuffield Council of Bioethics Working Party,[1] which exerts a strong influence on European biomedical scientists.

In Europe the most influential initiatives in the area of research ethics have been taken by the EU, with the directives on data protection and drug research, and by the Council of Europe through its above mentioned bioethics convention and additional protocols with their corresponding explanatory reports.

For European biomedical scientists, the following approach to ranking the many codes of research ethics can be advocated:

- Be familiar with your national legislation, if any, on biomedical research ethics. If your country has ratified the Council of Europe instruments and/or is an EU member state, the codes and directives issued by these two organisations will be consistent with your national legislation.

- If your country has not ratified the Council of Europe convention and related protocols (although it may be on the way to doing so), read the codes and follow the principles set out therein.

1.
Independent body established by the trustees of the Nuffield Foundation in 1991 to consider the ethical issues arising from developments in medicine and biology.

- Know the United Nations codes and guidelines and use them as advisory documents.

- Be familiar with the professional guidelines and declarations, such as the Declaration of Helsinki and let them guide you, but at the same time be aware of their shortcomings and inconsistencies with national law and with the European conventions and directives.

- When planning co-operative projects with developing countries, refer to the Nuffield Council of Bioethics Working Party's publications for guidance and advice.

From a global perspective, Europe has much to offer the world in the field of biomedical research ethics. Its cultural potential and its diversity of old and new democracies constitute an inspiring complexity, which can provide non-European countries with a range of values and democratic solutions that may be transposable to other countries wishing to create their own control systems, while ensuring respect for fundamental human rights as implemented in Europe, maintaining a balance between diversity and unity.

Bibliography

Bradford Hill, A., "Streptomycin treatment of pulmonary tuberculosis", *British Medical Journal*, 1948; 2: 796-820.

Fibiger, J., "On serum treatment of diphtheria", (original in Danish), *Hospitals Tidende*, 1898; 6: 309-25 and 337-50.

Lind, J., A *treatise of scurvy*, 1753.

Nuffield Council on Bioethics, *The ethics of research related to healthcare in developing countries*, London, Nuffield Foundation, 2002.

Paré, A. *Œuvres complètes*, ed. J.F. Malgaigne, Paris, 1840-41.

US National Institute for Health, *Guidelines for the Conduct of Research Involving Human Subjects at the National Institute of Health*, Bethesda, Maryland, NIH, 1995.

Council of Europe

(see http://www.coe.int/T/E/Legal_affairs/Legal_co-operation/ Bioethics/)

Convention for the Protection of Human Rights and Dignity of the Human Being with regard to the Application of Biology and Medicine: Convention on Human Rights and Biomedicine, Oviedo, Council of Europe, 1997.

Explanatory Report to the Convention for the Protection of Human Rights and Dignity of the Human Being with regard to the Application of Biology and Medicine: Convention on Human Rights and Biomedicine, Strasbourg, Council of Europe, 1997.

Draft Additional Protocol to the Convention on Human Rights and Biomedicine, on Biomedical Research, Strasbourg, CDBI, 2002.

Draft Explanatory Report to the Draft Additional Protocol to the Convention on Human Rights and Biomedicine, on Biomedical Research, Strasbourg, CDBI, 2002.

Relevant documentation and directives from other international organisations

Directive 2001/20/EC of the European Parliament and of the Council of 4 April 2001 on the approximation of the laws, regulations and administrative provisions of the member states relating to the implementation of good clinical practice in the conduct of clinical trials on medical products for human use, *Official Journal of the European Communities*, 2001; L 121/34.

World Medical Association, *The Declaration of Helsinki*, 5th revision, Edinburgh, WMA, 2000.

Unesco, *The Universal Declaration on the Human Genome and Human Rights: from theory to practice*, Paris, Unesco, 1999.

Council for International Organizations of Medical Sciences (CIOMS), *International Ethical Guidelines for Biomedical Research involving Human Subjects*, Geneva, CIOMS, 2001.

UNAIDS, *Guidelines Document: Ethical Considerations in HIV Preventive Vaccine Research*, UNAIDS, 2000.

Ethical dilemmas in research

Uses and abuses of biomedical research

by Jan Helge Solbakk

What are the norms and values underlying and guiding biomedical research? What is their nature? From where do they originate? To what extent can biomedical research be of help in protecting and promoting basic human rights? In what ways may biomedical research inadvertently violate humans or abuse human rights (Boyd et al. 1997)? These are some of the questions that will be addressed in the present chapter.

There are several possible ways of dealing with these questions. One way would be to focus attention on different forms of abusive biomedical research that have actually taken place throughout history and try to understand in what ways these dark events have influenced the shaping of ethical codes of medical research, in particular the emergence of human rights concepts and language relating to them. A prominent example of such a study is *The Nazi doctors and the Nuremberg Code: human rights in human experimentation* (Annas and Grodin 1992a).[1]

An alternative path would be to start with a definition of the purpose of biomedical research and try to identify situations where the attempt at achieving the legitimate aims of biomedical research could lead to abusive research and violation of human rights. The fruitfulness of this approach is evidenced in a report on medicine and human rights published by the British Medical Association (BMA 2001). According to the authors of this report, biomedical research is driven by two aims or desires (BMA 2001):

- scientific curiosity;

- the desire "... to benefit society by the systematic acquisition of useful, empirical knowledge".

"Research", the report continues, "is driven by a desire to understand the causes of disease or dysfunction and find effective methods of prevention and treatment. In extreme cases, however, even such humanitarian aims can be risky. The very

[1]
Although a human rights perspective is lacking, the study by Albert Jonsen represents a very instructive historical account of the shaping of ethical codes of medical research: Jonsen, A. R. "Experiments perilous: the ethics of research with human subjects", in Jonsen 1998.

potential for achieving tangible benefits can feed the temptation to press on beyond acceptable boundaries". In the report nine risk-factors for abusive research are identified (BMA 2001).

Factors that might lead to abusive research and violation of human rights:

- the power and influence of the researcher;
- the dependent situation of populations chosen as research subjects;
- the perception of a national necessity or government pressure to conduct research;
- the perception of an urgent and overriding scientific need;
- extreme detachment and lack of any sense of sympathy with the fate of research subjects;
- the perception that some people are expendable or already "terminal";
- the perception that some populations should be excluded from social concern; and
- secrecy.

We shall return to the issue of abusive biomedical research later in the chapter, but first it is necessary to get a clearer picture of the system of norms and values underlying and actually guiding biomedical research as well as a view of how this normative system relates to the framework of values witnessed in human rights documents.

The normative system of biomedical research

From a science-ethics point of view, medical research – as a representative form of scientific inquiry – may be defined as a systematic and socially organised:

1. search for;

2. acquisition of;

3. use or application of medical knowledge and insight brought forth by acts and activities involved in 1. and 2.[1]

1.
This definition is based on an action-theoretical conception of scientific research developed by the Norwegian philosopher and ethicist, Knut Erik Tranøy. For this, see Tranøy 1988b and Solbakk, J. H. 2001.

This action-theoretical conception of medical research pictures biomedical research as a "normative system", that is, as a finite and ordered set of norms and values for groups of people doing medical research. Furthermore, it broadens the scope to include not only the product and/or purpose of scientific inquiry but also the acts and activities – in sum the processes – that generate medical knowledge. Finally, it advocates a distinction between three different phases of scientific medical inquiry, the phases of :

- planning and search;
- acquisition;
- communication and use.

Before turning to an analysis of the different phases of scientific medical inquiry and of the possible forms of abuse and human rights violations that may emerge, a closer look at relevant core values in human rights documents seems justified.

Relevant core values in human rights documents

The first formulation of "an individual's right to health" within the framework of an "international human rights document" dates back to 1946 and the preamble to the constitution of the World Health Organization (BMA 2001). In subsequent human rights documents, this right to health is reiterated in various ways. For example, in Article 12 of the United Nations International Covenant on Social, Economic and Cultural Rights it is stated that "the enjoyment of the highest attainable standard of physical and mental health" is a right ascribed to everyone.

Furthermore, in Article 2 of the same covenant it is required of each state to take steps "to the maximum of its available resources, with a view to achieving progressively the full realisation of the rights" and Article 11 of the Council of Europe's Social Charter requires the parties to take measures to ensure the effective exercise of the right to protection of health, while in Article 3 of the Convention on Human Rights and Biomedicine of the Council of Europe, the parties' duty to provide within their own jurisdiction "equitable access to health care of appropriate quality" is underlined.

Another relevant article, and notably with explicit reference to biomedical research, is found in the second covenant generated from the Universal Declaration of Human Rights, that is, in the International Covenant on Civil and Political Rights. In Article 7 of this covenant, the core principle laid down in the Nuremberg Code – the principle of consent – figures prominently:

> No one shall be subjected to torture or to cruel, inhuman or degrading treatment or punishment. In particular, no one shall be subjected without his free consent to medical or scientific experimentation.

Among the many questions that are the subjects of extensive treatment and debate in the human rights literature, two seem to be of particular relevance within the present context of bio-medical research. First of all, the question of whether some human rights matter more than others. For example, should social and economic rights take priority over civil and political rights? And what about the right to development, which has been suggested as a third generation of human rights, besides the first generation rights of non-interference or "negative free-dom" laid down in the UN Covenant on Civil and Political Rights and the second generation of "positive liberty" wit-nessed in the International Covenant on Social, Economic and Cultural Rights? (BMA 2001).

The prudent stand advocated in the BMA report is that these rights should not be treated hierarchically, but according to their manifestation in the fabric of life, as rights which are interdependent and intertwined. The importance with respect to biomedical research of viewing the different categories of human rights as extensively interconnected will become evi-dent in subsequent paragraphs.

The second question that deserves attention within the present context is whether there is a need to establish an international tribunal to deal with researchers who have committed human rights violations in the name of biomedical research. According to the advocates of this proposal, international ethical codes and guidelines are necessary but not sufficient to deal with the problem of abusive research, because ethical guidelines lack the "... authority to judge and punish violators of international norms of human experimentation" (Annas and Grodin 1992b). We shall return to this question at the end of the chapter.

Norms and values underlying and guiding biomedical research

The search phase of biomedical research consists of two different kinds of systematic and socially organised searches for truth, each guided by different norms and values. First, there is the search in the science policy sense of the word. Second, there is the search in the sense of the design of individual research projects. While the latter kind of search "can only be organised and implemented by qualified researchers" (Tranøy 1996) under the guidance of the internal norms of science (Tranøy, 1988a)[1] the policy kind of search takes place under the guidance of external welfare norms and values such as utility, beneficence and equity, and linkage norms between the professional community of researchers and the community at large, such as fruitfulness and relevance. Consequently, this kind of search represents a shared responsibility between researchers and political decision-makers (Tranøy 1988b).

Also in the second and third phases of medical research – in the phases of acquisition and of communication and use of findings – the system of internal or methodological norms plays a paramount role. However, for obvious humanitarian reasons these methodological norms and values do not provide sufficient guidance with respect to the researchers' modes of behaviour towards their objects of research and towards subjects making use of their research findings (Jonas 1980):

> Experimentation was originally sanctioned by natural science. There it is performed on inanimate objects, and this raises no moral problems. But as soon as animate, feeling beings became the subjects of experiment, as they do in life sciences and especially in medical research, this innocence of the search for knowledge is lost and questions of conscience arise ...
>
> ...
>
> Human experimentation ... involves ultimate questions of personal dignity and sacrosanctity.

From these principal observations on moral conscience, moral concern and personal dignity, we shall now turn to a more concrete investigation of possible forms of abuse and human rights violations in the different phases of scientific medical inquiry.

1.
Among such norms we count for example originality, testability, inter-subjectivity, controllability, honesty, sincerity, exactitude, completeness, simplicity, order, coherence, consistency, dissimilarity, interestingness and objectivity. For a systematic view of this conglomerate of methodological norms and values, see Tranøy, K.E. 1988a.

Possible forms of abusive planning and search for medical data and knowledge

As to the question whether there exist forms of search or planning of research that deserve the label of "abusive search", the norm of freedom of inquiry favours a negative answer, in the sense that any prohibition in this early stage of planning and design of a research project seems unjustifiable. A tendency therefore, is to relegate the issue of forbidden knowledge to the second phase of research, the acquisition phase of scientific inquiry.[1] The observation made in the BMA report about the monstrous inequity in the world with respect to which/whose diseases are favoured in ongoing or planned research programmes hints, however, at the need for a different answer (BMA 2001):

> In 1996, ... it was estimated that approximately 56 billion US dollars was being spent annually on medical research and that at least 90% of this sum was devoted to the health needs of the richest 10% of the world's population. Therefore, the needs of 90% of the world's population have to be met from 10% of research funding. Infectious diseases, such as malaria, are responsible for more than half of the premature deaths among the poorest 20% of the world's population but only 7% of deaths among the richest 20%, who are more likely to suffer from conditions such as cerebro-vascular disease and ischaemic heart disease.

In the recently published WHO Report on Macroeconomics and Health: Investing in Health for Economic Development (World Health Organization 2001) this problem is dealt with in considerable detail, and a research strategy intended to reduce the gross inequity with respect to health and economic development is also proposed.

The underlying argument permeating the report gives support to the interdependency of human rights, previously argued in this chapter. Investment in health and essential health services in poor countries and in countries with low levels of income will not only reduce the disease burden in these countries, it will also generate economic growth and human flourishing. In turn, economic development will enable these countries to cope better themselves with their health problems.

1.
For a recent philosophical inquiry into the issue of "Forbidden Knowledge", see *The Monist* 1996, 79, 2.

Such a "global strategy for health" will, however, not be possible without a global medical science policy and research strategy that takes into account the particular research needs of these countries (WHO 2001). Four such research needs are identified in the report (WHO 2001):

- "operational research at the local level" to learn "what actually works, and why or why not";

- "a significant scaling up of financing for global R&D on the heavy disease burdens of the poor", such as HIV/Aids, malaria, tuberculosis, childhood infectious diseases and micronutrient deficiencies;

- "reproductive health", including research to block perinatal transmission of HIV;

- epidemiological research.

If these arguments prove to be valid, then it seems reasonable to draw the conclusion that countries not willing to contribute to this global research strategy for health and economic development should be held accountable for lending support to gross human rights violations in the name of biomedical research.

Possible forms of abusive acquisition of medical data and knowledge

"The fact that new information might be gained by an experiment", says W.K. Mariner, "does not, by itself, make the experiment ethically desirable or even justifiable. The Doctors' Trial at Nuremberg made abundantly clear, if it was not already obvious, that experiments can hurt people" (Mariner 1992). It thus also became clear why search for knowledge cannot be considered "the supreme value" in biomedical research (Mariner 1992) as well as making it clear why – in all international ethical guidelines and human rights documents addressing the issue of biomedical research – it is stated that the safety and welfare of human subjects should be the researcher's primary concern and take precedence over the interests of science and society.

From this it seems to follow that:

- biomedical research projects which involve a modest to high risk of serious harm to the research subject's life, health, privacy or dignity should be carefully monitored to prevent abusive forms of acquisition of medical knowledge slipping through. One possible exception may be projects aimed at studying seriously untreatable conditions, such as advanced cancer;

- research projects justified solely by reference to arguments about "national necessity" or "an urgent and over-riding scientific need" should be carefully monitored because of their potential for abuse and human rights violations.[1]

From the above-mentioned examples we shall move to research involving "vulnerable" persons as research subjects, namely people who due to "insufficient power, intelligence, education, resources, strength or other needed attributes" are in special need of attention to their situation of vulnerability which may necessitate, in some cases, protection of their interests by an ethics committee reviewing a research project for instance (CIOMS 2002). Examples of such groups are patients in emergency rooms, patients with incurable diseases, persons suffering from mental or behavioural disorders, residents of nursing homes, children, pregnant women, persons living in populations or communities with limited resources, ethnic and racial minority groups, unemployed or homeless people, members of the armed forces or police, refugees or displaced persons, and prisoners.[2]

To illustrate the potential for abusive research and human rights violations with respect to such groups of people, children and HIV-positive pregnant women living in developing countries will be used as examples.

A prevalent view in international ethical codes and guidelines regulating biomedical research, the Council of Europe's Convention on Human Rights and Biomedicine included,[3] is that children and other persons without the capacity to consent should be protected from being involved in research of no real and direct benefit to them, if it carries a risk that is greater than minimal. In spite of the good intentions behind such a safety measure, there are reasons to believe that it inadvertently may lead to situations where children in need of treatment will be exposed to greater therapeutic risks and dangers than necessary.

1.
For further elaboration of these arguments, including recent examples of such forms of abusive research, see BMA 2001: 211-213.

2.
For the examples listed and further examples, see CIOMS 2002.

3.
Convention on Human Rights and Biomedicine, Article 17, 2. ii.

This problem arises from the practice in non-therapeutic research of selecting adults instead of children as research participants and of developing new standards for approving paediatric use on the basis of extrapolation of data from studies in adults. This practice, the critics argue, has led to the paradoxical situation that children are often exposed to clinical decisions "without appropriate guidance from research" (Brody 1998). Consequently, diseased children are in danger of becoming "therapeutic orphans" (Brody 1998).

The "starting-with-children" approach that has been proposed, as an alternative to the classical "protectionist view" of extrapolation (Brody 1998), signals that although there may be morally relevant differences between children and adults, as subjects of medical research they should be treated as methodological equals. Therefore, in situations where the safety option of using adults is not available, children should not be systematically protected from participating in non-therapeutic research with a risk level that is greater than minimal, as such restrictions could lead to an infringement of their right to equitable access to health care of appropriate quality. This right is explicitly stated in Article 3 of the Convention on Human Rights and Biomedicine.

In 1997 a fierce standard of care debate broke out about the ethical acceptability of using placebo* as a comparative alternative to established effective treatment in trials conducted in developing countries for the purpose of preventing perinatal HIV-transmission.[1] The intervention, given the privileged status of "standard of care" in 1994, was the first randomised, controlled trial that successfully proved to be effective in reducing the incidence of human immunodeficiency virus (HIV) infection in pregnant women and their babies.

The drug employed in the trial with the complex name, Aids Clinical Trials Group (ACTG) Study 076 (Connor et al. 1994), was zidovudine. Its mode of administration was complex as well: oral administration to the pregnant women prior to giving birth, intravenous administration during labour and, after delivery, administration to the newborn infants as well. The intervention reduced the incidence of HIV infection by two thirds (Sperling et al. 1996).

Placebo:
a placebo is an inactive pill, liquid, or powder that has no treatment value. In clinical trials, experimental treatments are often compared with placebos to assess the treatment's effectiveness. In some studies, the participants in the control group will receive a placebo instead of an active drug or treatment.

1.
For an updated version of this controversy, see Solbakk, J.H. 2004.

Placebo-controlled trial: a method of investigation of drugs in which an inactive substance (the placebo) is given to one group of participants, while the drug being tested is given to another group. The results obtained in the two groups are then compared to see if the investigational treatment is more effective in treating the condition.

The main critique against the placebo-controlled trials* was that they were intentionally withholding a well-documented superior treatment (the "standard of care" regimen) from one group of pregnant women involved in the trials, namely, the control groups (Lurie et al. 1997; Angell 1997). The defenders of the use of placebo-controls, on the other hand, argued that the only way to proceed – in order to find a treatment that would be efficient as well as affordable for HIV-infected women living in these countries – would be to employ a placebo-controlled research design, since the answer to the question: "What reduction in rate of transmission would be of value to a developing country?, cannot be decided without reliable knowledge about the actual *in vivo* rate of perinatal transmission.[1]

The advocates of using a placebo-controlled study design also claimed that the main counterargument employed against the placebo controls – the immorality of withholding a well-documented superior treatment (the "standard of care" regimen) from the women in the control group – could be used to invalidate the use of "standard of care" as an ethically appropriate control regimen, as this would imply that the women in the intervention group were deliberately precluded from access to the best treatment option (Lie 1998; Lie 2002).

Furthermore, a trial using "standard of care" as control regimen not only suffers from the same sort – though not necessarily comparable in size – of ethical weakness identified in the trials using placebo controls, but it also violates the epistemological precondition deemed indispensable by all parties for conducting an ethical trial: the state of clinical equipoise (Lie 1998; Lie 2002).

1.
Quotation is from Lie 1998; p. 308.
See also Varmus, H., et Satcher, D. 1997, 337: 1003-5.

Finally, while the trials employing placebo could be acquitted of the usual charge of exploiting resource-poor countries, this is not the case with a trial using the "standard of care" regimen as an active control (Lie 1998; Lie 2002).

2.
For details about the kind of consensus that has been reached so far, see Solbakk 2004.

This controversy, which has not yet been fully settled,[2] provides a vivid illustration of the ethical complexities involved in acquiring scientifically sound knowledge of treatment options responsive to the medical needs and economic capabilities of people living in developing countries. On the other hand, the empirical evidence available so far lends little support to the view that the populations from which research subjects have been drawn, have actually benefited much from the results of the contested trials (Steinbrook 2002a; Steinbrook 2002b).

From a human rights perspective this represents a disturbing example of lack of implementation and use of research findings that are vital to the health needs of those communities that have hosted the trials. Consequently, it may also be argued that these communities have become victims of therapeutic neglect and abuse, in the sense that their rights to access to treatment of appropriate quality have not been met.

Possible forms of abusive communication and use of medical data and knowledge

A final form of abuse belongs to the phase of communication and application (or use) of biomedical research findings, and it deserves attention within the present context of human rights. This is abuse related to the notion of secrecy. "There is neither a legal nor a moral obligation to state, publish or communicate whatever we know simply for the reason that we know it." (Tranøy 1996).

In research, on the other hand, the situation is quite different. That is, regardless of what field of research we are dealing with, scientific statements should not be concealed; they should be made public, so that their scientific validity can be tested and checked by other scientists. Only in this way can a researcher comply with the norms and values of good scientific conduct. That is also the reason why a sponsor's wish to keep secret the results of a research project – or to introduce certain restrictions on the researcher's right to publish results from the project – represents one of the great moral challenges on the use or application-level of biomedical research (Tranøy 1996).

Lack of transparency and secrecy should therefore always be carefully monitored by the responsible health authorities, because these are factors that have proved to be present in all forms of abuse and human rights violations in the name of biomedical research (BMA 2001).

A question previously posed, but not yet answered, will now finally be addressed, viz. whether there is a need for establishing an international tribunal to deal with researchers who have committed human rights violations in the name of biomedical research.

An underlying argument throughout this chapter has been that there is a need for some sort of international instrument or forum to bring evidence to the global community about the gross inequity in the world with respect to which/whose diseases are favoured in ongoing or planned research programmes and with the responsibility of developing a medical science policy and research strategies aimed at meeting the particular research needs of poor and low-income countries. Such a forum could also serve as an instrument to monitor ongoing research, in order to safeguard communities and populations in those countries from being exploited in the name of biomedical research and medical treatment.

Personally, I believe a forum aimed at uncovering the political and structural deficiencies generating inequities in the world with respect to health-related research and treatment would be more necessary than an international tribunal aimed at targeting human rights violations committed by individual researchers. For such a forum to be able to function in a proactive way, close collaboration with national health authorities – as well as with international bodies such as the UN, WHO and the World Bank – would be important.

In the WHO report on macroeconomics and health previously referred to (WHO 2001), two proposals are put forward that seem to prefigure the idea of creating such a forum – first, the establishment of National Commissions on Macroeconomics and Health in developing countries, with the tasks of:

• assessing "national health priorities";

• proposing strategies for the "coverage of essential health services";

- preparing "an epidemiological baseline, quantified operational targets, and a medium-term financing plan".

and second, the creation of a Global Health Research Fund (GHRF) to "... support basic and applied biomedical and health sciences research on the health problems affecting the world's poor and on the health systems and policies needed to address them" (WHO 2001).

An endorsement of those proposals by the international political community and commitment on the part of those countries capable of contributing resources to such a research fund would be powerful signals to the world of biomedical research that human rights matter. The creation of a global forum of inequities in health-related research could make this message come true.

Bibliography

Angell, M. 1997. "The ethics of clinical research in the Third World" (editorial), *New England Journal of Medicine*, 337; 12: 847-9.

Annas, G.J. and Grodin, M.A. (eds) 1992a. *The Nazi Doctors and the Nuremberg Code : Human Rights in Human Experimentation*, Oxford; Oxford University Press.

Annas, G.J. and Grodin, M.A. 1992b. "Where do we go from here ?" in Annas and Grodin, 1992a, p. 313.

British Medical Association 2001. "Research and experimentation on humans", *The Medical Profession and Human Rights : Handbook for a changing agenda*, BMA.

Boyd, K., Higgs, R. and Pinching A. (eds) 1997. *The New Dictionary of Medical Ethics*, BMJ Publishing, p. 126.

Brody, B. 1998. *The Ethics of Biomedical Research : An International Perspective*, New York, Oxford University Press, p. 177.

CIOMS (Council for International Organizations of Medical Sciences) 2002, "Commentary on Guideline 13", *International Ethical Guidelines for Biomedical Research Involving Human Subjects*, Geneva.

Connor, E.M., Sperling, R.S., Gelbert, R. et al. 1994. "Reduction of maternal-infant transmission of human immunodeficiency virus type 1 with zidovudine treatment", *New England Journal of Medicine*, 331: 1173-80.

Jonas, H., 1980. "Philosophical reflections on experimenting with human subjects", in H. Jonas, *Philosophical Essays : From Current Creed to Technological Man*, Chicago, University of Chicago Press, pp. 105-35.

Jones, A.J.I. (ed.), K.E. Tranøy, 1988. *The Moral Import of Science : Essays on normative theory, scientific activity and Wittgenstein*, Norway Sigma Forlag, pp. 111-21.

Jonsen, A.R. 1998. "Experiments perilous : the ethics of research with human subjects", in Jonsen, A.R., 1998. *The birth of bioethics*, New York, Oxford University Press, pp. 125-65.

Lie, R.K. 1998. "Ethics of placebo-controlled trials in developing countries", *Bioethics*, 12 ; 4 : 307-11, at p. 308.

Lie, R.K. 2002. "The HIV perinatal transmission studies" in Lie, R.K., Schotsmans, P.T., Hansen, B., and Meulenbergs, T. (eds), *Healthy Thoughts : European Perspectives on Health Care Ethics*, Leuven, Peeters, pp. 189-206, at p. 198.

Lurie, P. and Wolfe, S.M. 1997. "Unethical trials of interventions to reduce perinatal transmission of the human immunodeficiency virus in developing countries", *New England Journal of Medicine*, 337; 12: 853-6.

Mariner, W.K. 1992. "AIDS research and the Nuremberg Code" in Annas and Grodin, 1992a, pp. 286-303.

Solbakk, J.H. 1998. "The concept of goodness in medical research : an action theoretic approach" in D. Weisstub (ed.), 1998, *Research on Human Subjects : Ethics, Law and Social Policy*, Oxford, Pergamon, pp. 73-87.

Solbakk, J.H. 2001. "Le concept du bien dans la recherche médical : une approche fondée sur une théorie de l'action" in Weisstub, D.N., Mormont, C. and Herve, C. (eds), 2001, L'Éthique de l'expérimentation sur les être humains, Vol. 1: *Réflexion philosophiques et historiques*, Paris, L'Harmattan, pp. 123-42 (French version).

Solbakk, J.H. 2004. "Use and abuse of empirical knowledge in contemporary bioethics : a critical analysis of empirical arguments employed in the controversy surrounding studies of maternal-fetal HIV-transmission and HIV-prevention in developing countries", *Journal of Medicine, Healthcare and Philosophy*, Volume 7, No. 1; 2004: 5-16.

Sperling, R.S., Shapiro, D.E., Coombs, R.W. et al. 1996. "Maternal viral load, zidovudine treatment, and the risk of transmission of human immunodeficiency virus type 1 from mother to infant", *New England Journal of Medicine*, 335; 22: 1621-9.

Steinbrook, R. 2002a. "Preventing HIV infection in children", *New England Journal of Medicine*, 346; 24: 1842-3.

Steinbrook, R. 2002b. "Beyond Barcelona – the global responses to HIV", *New England Journal of Medicine*, 347; 8: 553-4.

Tranøy, K.E. 1988a. "The foundations of cognitive activity: an historical and systematic sketch" in Jones, A.J.I. (ed.), Tranøy, K.E., 1988, pp. 121-36.

Tranøy, K.E. 1988b. "Science and ethics. Some of the main principles and problems", in Jones, A.J.I. Ed., Tranøy, K.E., 1988, pp. 111-21.

Tranøy, K.E. 1996. "Ethical problems of scientific research: an action-theoretic approach", *The Monist*, 79; 2: 183-96.

Varmus, H. and Satcher, D. 1997. "Ethical complexities of conducting research in developing countries", *New England Journal of Medicine*, 1997, 337; 14: 1003-5.

World Health Organization 2001. Report of the Commission on Macroeconomics and Health, Geneva.

Selection and recruitment of participants: European standards

by Herman Nys

The selection and recruitment of participants in biomedical research has been the subject of many regulations at international, European and national level. Internationally, the Helsinki Declaration contains recommendations guiding physicians involved in biomedical research on human subjects. It was adopted and subsequently amended on various occasions by the World Medical Association. In 1996, also at international level, the International Conference on Harmonisation (ICH) issued a set of Good Clinical Practices that provide a unified standard for the European Union, Japan and the United States.

Both the Helsinki Declaration and the ICH Good Clinical Practices influenced European Parliament and Council Directive 2001/20/EC of 4 April 2001 on the approximation of the laws, regulations and administrative provisions of the EU member states relating to the implementation of good clinical practice in the conduct of clinical trials on medicinal products for human use. Recital 2 of the directive expressly refers to the Helsinki Declaration as an accepted basis for the conduct of clinical trials. It is, however, remarkable that reference should be made to the 1996 version of the Helsinki Declaration and not to the 2000 version.

The same recital also refers to "the protection of human rights and the dignity of the human being". These words reflect the title of the European Convention for the Protection of Human Rights and Dignity of the Human Being with regard to the Application of Biology and Medicine. One wonders why this convention was not considered worthy of an explicit reference.

The international and European regulations all contain rules governing the protection of participants in biomedical research. These rules not only offer protection but, for those undertaking such research (sponsors, investigators and so on), they provide guidance in the process of selecting and recruiting participants. This chapter is mainly concerned with the rules and standards governing the protection of participants in biomedical research.

It is not my intention to provide a detailed analysis of these rules and standards by comparing the texts and documents that have already been mentioned. Rather, I want to indicate the main points relating to the selection and recruitment of participants in biological research in the EU directive on good clinical practice in clinical trials. The references in brackets in the text are to the directive. For a more detailed discussion of the rules in the European Convention on Human Rights and Biomedicine and the Additional Protocol on Biomedical Research, I would refer you to the chapter by Pēteris Zilgalvis in this publication.

The protection of participants in biomedical research: general rules

Scientific soundness

The first condition to be met, before starting to recruit and select participants for biomedical research, is the scientific soundness of the research project. This condition covers two aspects: firstly, the research project itself and, secondly, the staff, equipment, infrastructure and so on. Regarding the first aspect, the so-called research protocol must describe the objective(s), design, methodology and organisation of the research project, along with the statistical considerations. The protocol itself must be based on reliable laboratory and animal research. Furthermore, the research project must be feasible and yield relevant information. Regarding the second aspect, the investigator carrying out/supervising the research must be a medical doctor, because of the scientific background and the experience in patient care that the research requires. Each staff member must have suitable educational qualifications, training and experience.

Risk/benefit ratio

The second condition is that an appropriate risk/benefit ratio must be respected. This means that, before a research project is initiated, the foreseeable risks (the term is tautologous, because an event that is not foreseeable as medical science stands is not a risk) and inconvenience should be weighed against the anticipated benefit for the individual participant, other present and future patients and society at large (Article 3.2 (a)). A research project should be initiated and continued only if the anticipated therapeutic and public health benefits justify the risks.

This evaluation is not a one-off risk assessment but an ongoing process : newly available information must be considered. This is the only way of ensuring that the investigator can end the research project prematurely if its continuation will lead to damage, death or other serious complications, that is, when the risks exceed the benefits. If there is prior reason to suppose that death or disability will occur, the research project must not be carried out. Most international documents mention this prohibition.

Informed consent

The "golden standard" of biomedical research is the free, voluntary, express and informed consent of the selected participant (or his or her representative – see below). In a prior interview with the investigator or a member of the investigating team, potential participants must have had the opportunity to understand the objectives and risks of the research project, the inconvenience it entails and the conditions under which it is to be conducted. They must also be informed of their right to withdraw from the research project at any time (Article 3.2 (b)).

Participants must give their written consent after being informed of the nature, significance, implications and risks of the research project. If the participant is capable of giving his or her consent but unable to write (because of a physical disability, for instance), oral consent in the presence of at least one independent witness may be given (Article 3.2 (d)). Participants may, without any resulting detriment, withdraw from the research project at any time by revoking the informed consent orally or in writing (Article 3.2 (e)). Participants must be provided with a contact point where they may obtain further information (Article 3.4).

Medical care, insurance and indemnity

Any medical care given to, and medical decisions made on behalf of, participants during the research project are the responsibility of an appropriately qualified doctor (Article 3.3). Provision must be made for insurance or indemnity to cover the liability of the investigator and the sponsor of the project (Article 3.2 (f)).

The protection of minor participants in biomedical research: specific rules

Apart from the general rules that govern the protection of participants in biomedical research, the following specific rules must be observed when minors are recruited and selected for such research.

Subsidiarity

A fundamental rule in biomedical research is the subsidiarity rule. It means that, in principle, only persons who can give their free and informed consent may be included in a research project. If the project can take place with healthy, independent subjects capable of giving their consent, no subjects belonging to so-called "vulnerable persons", such as minors, may be included in the project.

From this rule and the rule of scientific soundness, it follows that research on minors has to be essential for the purposes of validating data obtained in research projects on persons able to give their informed consent or by other research methods (Article 4 (e)). Subsidiarity also implies that research should either directly relate to a clinical condition from which the minor concerned suffers or be of such a nature that it can be carried out only on minors.

Direct benefit for the participant

Healthy persons capable of giving their informed consent may participate in biomedical research without any anticipated direct benefit to them (so-called "non-therapeutic biomedical research"). This is not the case for minor participants: some direct benefit has to be obtained from the research project for the minors themselves or for the group to which they belong by virtue of their age and/or illness (Article 4 (e)).

Informed consent of the parents

The informed consent of the parents of the minor or his or her legal representative must be obtained. This consent may be revoked at any time, without detriment to the minor (Article 4 (a)).

Considering the wishes of the minor

The explicit wish of a minor who is capable of forming an opinion to refuse participation or to be withdrawn from the research project at any time must be considered by the investigator (Article 4 (c)).

Minimal risk and minimal discomfort

A research project in which minors are participating has to be designed to minimise pain, discomfort, fear and any other foreseeable risk. Both the risk threshold and the degree of distress have to be specially defined and constantly monitored (Article 4 (g)).

The protection of incapacitated adults participating in biomedical research: specific rules

In the case of adults incapable of giving their informed consent, all the relevant requirements listed for persons capable of giving such consent are applicable. In addition to these requirements, inclusion in biomedical research of incapacitated adults is allowed only if the following conditions are also met.

Subsidiarity

The research project is essential to validate data obtained in research projects on persons able to give their informed consent or by other research methods (Article 5 (e)).

Direct benefit for the participant

There are grounds for expecting the research project to produce a direct benefit to the participant (Article 5 (i)). Moreover, the project relates directly to a life-threatening or debilitating clinical condition from which the incapacitated adult concerned suffers. In other words, only research that is of direct benefit to the participant, and the condition from which he or she is suffering, is allowed (Article 5 (e)). A benefit for the group of patients that the participant belongs to is not sufficient. However, there is no consensus on this narrow viewpoint and, for instance, Article 17 of the Convention on Human Rights and Biomedicine allows more leeway in this respect (see the chapter by Zilgalvis).

Informed consent

The informed consent of the legal representative must have been obtained (Article 5 (a)). In many countries it is unclear who this representative is, because the large majority of incapacitated adults do not come under a specific system of protection. In practice, the next of kin are often considered as the representatives of the incapacitated person, but from a strict legal point of view this practice is disputed. A question that arises in this respect is how binding an advance decision by an adult who has since become incapable of giving his or her consent is.

In theory, one may consent in advance to a research project. But it is debatable whether such informed consent can ever be specific enough to be valid. Was the person, at the time when he or she consented in advance, really in a position to evaluate all the risks? That may be so, but it would seem to be rather exceptional for it to be the case. Such informed consent may of course influence the decision of the legal representative, but it cannot be considered as binding.

The situation is different in the case of an informed refusal. One has to accept that a competent adult may refuse in advance, either in specific or in general terms, to participate in a particular research project or in any research project whatsoever. This advance refusal has to be respected by the legal representative and the investigators. Although from a moral point of view one may defend an obligation to participate in biomedical research, such an obligation does not exist in legal terms.

The consent of the representative may be revoked at any time, without detriment to the subject (Article 5 (a)).

Considering the wishes of an incapacitated adult

A person not able to give his or her informed legal consent must have received information according to his or her capacity of understanding regarding the trial, the risks and the benefits. The explicit wish of a subject who is capable of forming an opinion and assessing this information to refuse participation in, or to be withdrawn from, the medical research project at any time must be considered by the investigator (Article 5 (b) and (c)).

Minimal risk and minimal discomfort

A research project in which incapacitated adults are participating has to be designed to minimise pain, discomfort, fear and any foreseeable risks in relation to the disease and developmental stage; both the risk threshold and the degree of distress must be specially defined and constantly monitored (Article 5 (f)).

A favourable opinion from an ethics committee

A most important step in the recruitment and selection of participants for biomedical research is the evaluation of the research protocol by an ethics committee. There is a broad consensus in all international documents and declarations regarding this requirement, although there may be differences as to the membership of such committees, their competence and so on.

Favourable opinion before the start of a research project

A biomedical research project in which human subjects are to participate may not start until an ethics committee has issued a favourable opinion (Article 6.2.).When the subjects involved are minors or incapacitated adults, the ethics committee must endorse the protocol (Article 4 (h) and Article 5 (g)).

The favourable opinion requirement is legally binding in two senses. Firstly, if the opinion is unfavourable a biomedical research project cannot be started. Secondly, a protocol that has received a favourable opinion may not be changed during the research project. If, after the commencement of the project, the sponsor of the project wants to make an amendment to the protocol, the sponsor must inform the ethics committee, which has to give its opinion on the proposed amendment. If that opinion is unfavourable, the amendment to the protocol may not be implemented (Article 10 (a)).

Responsibilities of an ethics committee

It is the responsibility of an ethics committee to protect the rights, safety and well-being of human subjects involved in the research project and to provide public assurance of that protection (Article 2 (k)). To this end, the ethics committee has to consider in particular the relevance of the research project and its design; whether the balance of the anticipated benefits and the risks is satisfactory; the research project protocol; the suitability of the investigator and the supporting staff; the investigator's information brochure and the quality of the facilities. It also has to consider the adequacy and completeness of the written information to be given to the participants and the justification for research on persons incapable of giving informed consent.

Further, it has to evaluate the provision made for indemnity or compensation in the event of injury or death attributable to the research project and insurance or indemnity to cover the liability of the investigator or the sponsor. One particularly important responsibility as regards the recruitment and selection of participants is that the ethics committee has to consider the amounts and, where appropriate, the arrangements for rewarding or compensating investigators and participants (Article 6.3 (a) to (j)). With regard to minors and incapacitated adults, no incentives or financial inducements may be given, except compensation (Article 4 (d) and Article 5 (d)). Finally, an ethics committee has to consider the arrangements for the recruitment of subjects (Article 6.3. (k)).

Membership of an ethics committee

An ethics committee must consist of healthcare professionals and non-medical members (Article 2 (k)). When minors are involved as participants, the ethics committee should have paediatric expertise or should seek advice on clinical, ethical and psychosocial problems in the field of paediatrics (Article 4 (h)). When incapacitated adults are involved, the ethics committee should have expertise in the relevant disease and the patient population concerned or seek advice on clinical, ethical and psychosocial questions in the field of the relevant disease and population concerned (Article 5 (g)).

Placebo: its action and place in health research

by Andrzej Górski

Appendectomy: surgical removal of the vermiform appendix. This procedure is normally performed as an emergency procedure when the patient is suffering from acute appendicitis.

Ongoing progress in biomedical research and technology promises to offer a substantial improvement in the quality of people's lives as well as generate important economic benefits. However, those advances can also give rise to difficult ethical questions (Gerold 2004; Koski and Nightingale 2001; Shapiro and Meslin 2001). These problems are especially relevant in clinical research with its inherent tensions between the ethical values of pursuing rigorous science and of protecting participants from undue harm.

In this context, it should be emphasised that the ethics of clinical research is not tantamount to the ethics of clinical care: the primary goal of clinical trials is to advance medical knowledge, not to promote patients' best medical interests (which may be compromised by their exposure to risks not necessarily outweighed by known potential medical benefits (Emanuel et al. 2000; Horng and Miller 2002). Therefore, the biomedical research community, international institutions and organisations, and the public have been engaged in a re-examination of the ethical and responsible conduct of research involving human participants for many years.

Among the issues that have received great attention and have been a matter of intense public controversy has been the role and place of placebo in health research. The course of discussion and the controversies surrounding the subject of placebo are not unique in the history of bioethics. For example, the world's first laparoscopic appendectomy* was performed in 1980 by Kurt Semm in Kiel, Germany. The president of the German Surgical Society called for his suspension, and the relevant paper was rejected because the procedure was at that time considered unethical (Tuffs 2003). Today laparoscopic surgery is an accepted treatment.

Hippocrates (c.460-438 BC), regarded as the father of medicine and author of the Hippocratic oath, was probably the first to notice that medical attention itself can make a patient feel better and induce an effect now referred to as placebo:

> The patient, though conscious his condition is perilous, may recover his health simply through his contentment with the goodness of the physician. (Box 2004)

Placebo can be defined as "a dummy treatment administered to the control group in a controlled clinical trial in order that the specific and non-specific effects of the experimental treatment can be distinguished" (see On-Line Medical Dictionary: http://cancerweb.ncl.ac.uk/omd).

A placebo can be pharmacological (a tablet), physical (a manipulation), or psychological (a conversation, for example, being a part of medical attention). An in-depth analysis by Hrobjartsson et al. (Hrobjartsson and Gotzsche 2001) suggests that there is little evidence that placebo has significant beneficial effects; therefore, there is no justification for its use outside clinical trials. However, as pointed out by Emanuel and Miller (2001), patients given "no treatment" (group in addition to a placebo group) were in fact receiving clinical attention that may have contributed to observed improvements. Thus, clinical attention may be responsible for the placebo effect, and clinical trials involving such attention in fact test whether the experimental treatment is better than this attention, not whether it is better than nothing.

Furthermore, it has been recognised that a placebo can cause the noxious or distressing effects referred to as the nocebo phenomenon (Latin: *I will harm,* as opposed to placebo: *I will please*). In fact, approximately a quarter of patients on placebo report such adverse side-effects: drowsiness, nausea, fatigue and insomnia, and their incidence may even exceed the incidence of side-effects in patients taking the active drug (Barsky et al. 2004).

Evidently, any treatment (pharmacological, physical or psychological) may have a beneficial effect, since patients perceive themselves as being treated. This in turn activates biochemical mechanisms, which can lead to clinical improvement. Therefore, the placebo effect is basically related to the expectation of clinical benefit (a phenomenon that may explain placebo responses obtained in treating pain, depression and Parkinson's disease).

Recent studies indicate that placebo is related to the limbic system of the brain and may be mediated by dopamine – a chemical that transmits pleasure signals to the brain and is therefore responsible for a placebo's beneficial effects. On the other hand, the ability of placebo to alleviate pain (placebo analgesia) is mediated by endogenous opioids in the context of a patient's expectations. This results from the doctor–patient interaction, but even the mere presence of medical staff may be important. Aside from dopamine and opioids, other endogenous mediators may also be involved, such as cholecystokinin (de la Fuente-Fernandez and Stoessl 2004).

Placebo-controlled clinical trials (PCT) have been criticised since their initiation: in 1931, sanocrisin was compared with distilled water in the treatment of tuberculosis (Emanuel and Miller 2001). However, some scientists and doctors believe that important negative consequences would result from uniformly prohibiting PCT and allowing only active controls (a therapy effective in treating a given condition) and suggest that it would be difficult if not impossible to identify new treatments representing a major advance without studying them in such trials.

The "active-control orthodoxy" group cites the Declaration of Helsinki, which elevates concern for the health and rights of individual patients above concern for society, for future patients, and for science :

> The benefits, risks, burdens, and effectiveness of a new method should be tested against those of the best current prophylactic, diagnostic and therapeutic methods. This does not exclude the use of placebo, or no treatment, in studies where no proven prophylactic, diagnostic or therapeutic method exists. (Declaration of Helsinki, Provision 29)

If patients are assigned a placebo instead of a therapy effective in treating their condition, placebo opponents argue that such trials are in breach of the declaration, which – in their judgement – proscribes the use of a placebo as a control, when a proven therapeutic method exists. They also argue that the use of a placebo allows for approval of drugs of undetermined efficacy, as the scientific benefit of PCT is illusory (the effect of a new drug appears large and may be statistically significant even in a small study, so the evaluation of results is subject to statistical errors).

The 5th revision of the Declaration of Helsinki (October 2000), containing the above-mentioned rigid standards in its placebo control guidelines ignited much controversy and even criticism (Forster et al. 2001). Consequently, in 2002 a note of clarification was added that reads:

> The WMA hereby reaffirms its position that extreme care must be taken in making use of placebo-controlled trials and that in general this methodology should only be used in the absence of existing proven therapy. However, a placebo-controlled trial may be ethically acceptable, even if proven therapy is available, under the following circumstances:
>
> • where, for compelling and scientifically sound methodological reasons, its use is necessary to determine the efficacy or safety of a prophylactic, diagnostic or therapeutic method; or
> • where a prophylactic, diagnostic or therapeutic method is being investigated for a minor condition and the patients who receive placebo will not be subject to any additional risk or serious irreversible harm.

Clear guidance on the matter of placebo is provided by the Council for International Organizations of Medical Sciences (CIOMS), which uses the term "established effective intervention" instead of "best current therapeutic method" or "proven therapeutic method" (Helsinki Declaration), as the active comparator that is ethically preferred in controlled clinical trials. This avoids problems when there may be lack of agreement in the medical world as to which method is the best (Idanpaan-Heikkila and Fluss 2004).

CIOMS Guideline 11 (International Ethical Guidelines for Biomedical Research Involving Human Subjects) states that:

> As a general rule, research subjects in the control group of a trial of a diagnostic, therapeutic, or preventive intervention should receive an established effective intervention. In some circumstances, it may be ethically acceptable to use an alternative comparator, such as placebo or "no treatment".
>
> Placebo may be used:
>
> • when there is no established effective intervention;
> • when withholding an established effective intervention would expose subjects to, at most, temporary discomfort or delay in relief of symptoms;
> • when use of an established effective intervention as comparator would not yield scientifically reliable results and use of placebo would not add any risk of serious or irreversible harm to the subjects.

The Council of Europe is recognised for its leading role in bioethics and its focus on human rights. Therefore, it would be important to present its views on placebo as expressed by the Additional Protocol on Biomedical Research, supplementary to the Convention on Human Rights and Biomedicine, adopted by the Council of Europe's Committee of Ministers in June 2004. It addresses such issues as consent, the protection of persons not able to consent to research, and research ethics committees, and it is the first internationally binding legal instrument covering the field of biomedical research.

Article 23 states :

> the use of placebo is permissible where there are no methods of proven effectiveness, or where withdrawal or withholding of such methods does not present unacceptable risk or burden.

Thus, as pointed out by Zilgalvis (Zilgalvis 2004), the protocol's definition represents a kind of "middle ground" proposing a realistic balance between the proponents and opponents of the use of placebo.

In 2000 the European Parliament, the Council and the Commission proclaimed the Charter of Fundamental Rights, whose Article 3 addresses important issues of medicine and biology, such as free and informed consent, the prohibition of selling the human body and its parts, and the prohibition of reproductive cloning.

Recently, the Commission has also formulated its stance on the issue of placebo, as seen from the perspective of the European Group on Ethics in Science and Technology (EGE). Established in 1990, the EGE advises the President of the Commission on questions related to bioethics and biotechnology. It has issued its Opinion No. 17 (February 2003), stating that :

> placebo-controlled trials may be acceptable when, for example, the primary goal of the clinical trial is to try to simplify or decrease the costs of treatment for countries where the standard treatment is not available for logistic reasons or inaccessible because of the cost.

This opinion was not accepted unanimously, as two members dissented from that conclusion, believing that it might allow for double standards for research in wealthy and poorer countries.

Furthermore, European Commission Directive 2001/83/EC supports the Declaration of Helsinki:

> All clinical trials shall be carried out in accordance with the ethical principles laid down in the current revision of the Declaration of Helsinki ...

> In general, clinical trials shall be done as controlled clinical trials, and if possible, randomized ... any other design shall be justified.

It also adds:

> The control treatment of the trials vary from case to case and also will depend on ethical considerations...

> It may, in some instances, be more pertinent to compare the efficacy of a new medicinal product with that of an established medicinal product of proven therapeutic value rather than with the effect of placebo.

In addition, the European Agency for the Evaluation of Medicinal Products (EMEA) considers the issues of Good Clinical Practice (GCP) – including ethical issues such as informed consent, approval by ethics committees and the like – when granting marketing authorisation. When problems are detected, the EMEA can refuse or withdraw marketing authorisation and advise the Commission. In June 2001, the EMEA adopted a position statement on "The use of placebo in clinical trials with regard to the revised Declaration of Helsinki". Affirming that the Declaration remains a vital expression of medical ethics whose aims deserve unanimous support, it also states that:

> Forbidding placebo-controlled trials in therapeutic areas where there are proven prophylactic, diagnostic and therapeutic methods would preclude obtaining reliable scientific evidence for the evaluation of new medical products, and be contrary to public health interest ... Provided that the conditions that ensure the ethical nature of placebo-controlled trials are clearly understood and implemented, it is the position of the EMEA that continued availability of placebo-controlled trials is necessary to satisfy public health needs. (www.wma.net/e/ethicsunit)

As summarised above, discussions about the use of placebo have been particularly intense and polarised, and have perhaps reached an impasse: while some approve of only a very limited role for placebo and suggest that the need for them is over-stated (Michels and Rothman 2003), others have endorsed more widespread use (Temple and Meyer 2003; Temple et al. 2000; Ellenberg et al. 2000; Singer 2004). As recently pointed out by Sugarman (2004) this debate has centred on rhetoric rather than on analysing relevant empirical findings.

Therefore, to assess whether PCT carry an increased risk of harm, the results of eighty such trials in patients with mild to moderate hypertension were evaluated. It appears that a short-term PCT in those patients does not pose an elevated risk. Likewise, an analysis of such trials involving more than 19 000 patients with depression did not reveal differences between active treatment and placebo (Sugarman 2004).

Obviously, physicians may be key players in the enrolment of patients in clinical trials, and the attitude of the public may also contribute to the place of placebo in health research today. A nationwide mailed survey in the USA has revealed that physicians preferred active-control trials and believed that they are superior to placebo-controlled ones (on the basis that they are more likely to lead to a patient benefit and public benefit and are less likely to expose patients to risks). However, it is unclear whether the respondents had a thorough understanding and knowledge of the complexity of the problem; for example, the number of currently used treatments that have not been validated (Sugarman 2004).

To break the present impasse, national consultations were conducted in Canada under the aegis of Health Canada and the Canadian Institutes of Health Research (the National Placebo Initiative). Potential participants were identified from five major cities in Canada and the consultations were performed using the process of deliberative dialogue (involving video sessions presenting an overview of the current use of PCT, the policies guiding its use, and the assessment of PCT by ethics commissions (institutional review boards).

Participants supported the use of placebo as an important tool in medical research and the advancement of science and thought that it constitutes a necessary and valid part of developing and testing new treatments. At the same time, they emphasised the necessity of full informed consent and minimising potential conflicts of interest as well as the fact that placebo use should be selective (believing its applicability should decrease when there is an increasing risk for the trial patients). Some participants believed that patient autonomy should be the highest priority and the first consideration: whenever possible, considering the patient as a partner and giving patients the choice to participate in PCT.

Interestingly, the participants accepted that there might be some risk involved in such trials, but valued the idea of their country as a leader in medical research and wished to avoid the negative effects of overly restrictive placebo policy. Even though the total number of participants of the study was relatively small, it should be emphasised that for the first time public consultations were conducted nationally on such a key bioethical issue as placebo (Huston 2004).

The outcome of the Canadian National Placebo Initiative is similar to the views expressed by Dr Temple, Director of the Office of Medical Policy, Center for Drug Evaluation and Research of the Federal Drug Administration (FDA) in the USA, who believes that there is a continued need for placebo even when there is effective therapy (Temple and Meyer 2003). In his important paper published in 2000 he considers the ethical concerns about the use of placebo controls and emphasises the limited ability of active-control trials to establish efficacy of new therapies.

Temple and Ellenberg believe that in conditions in which forgoing therapy carries no important risk, the participation of patients in PCT seems appropriate and ethical – as long as patients are fully informed (Temple and Ellenberg 2000). Moreover, they point out that not all placebo studies leave patients untreated, as it is frequently possible to provide standard therapy while carrying out a superiority study (intended to demonstrate an advantage of a treatment regimen over the control – for example using an "add-on" design in which all patients are given standard therapy and are also randomly assigned to receive either a new agent or a placebo). Also, it has to be kept in mind that even FDA approval does not exclude the continuation of PCT when drug efficacy or its side-effects are still in doubt.

Furthermore, the authors emphasise a longer-range rationale for patient participation in a PCT. The trial participants themselves may benefit in the future regardless of their assigned treatment in the initial trial (active drug versus placebo). Thus, it could be argued that forgoing active treatment for the short term may improve long-term chances of successful treatment. In addition, it must be recognised that a new treatment may represent a major advance without being more effective than alternatives, which is another argument for the negative consequences of uniformly prohibiting PCT (Temple et al. 2000; Ellenberg et al. 2000).

In the United Kingdom, PCT seem to be less popular than in the USA. The UK Medical Research Council, which supports high-quality research aimed at improving human health, is currently funding 190 trials, and only 20 of them are PCT (Box 2004) – the details are available on the website http ://www.controlled-trials.com/. However, elsewhere in Europe the opinions on PCT also vary.

Thus, placebo is valued at the prestigious University of Vienna: the head of its ethics commission believes that waiving the PCT altogether may be as unethical as its unjustified use. In his lecture during the placebo conference in April 2003 in Warsaw, he presented examples of clinical trials which support the importance of a placebo group, without which one might never have learned about the uselessness of the suggested and subsequently tested clinical treatment. This is another important argument for the consideration of placebo, since scientifically invalid research cannot be ethical, no matter how favourable the risk/benefit ratio for the study participants may be. The author rightly points out that placebo is not an ideal solution, but clinical research cannot provide ideal solutions, only the best possible ones (Singer 2004).

Within the subject of placebo, nothing could be more provoca-tive and challenging than the use of placebo in surgery. It is known that surgical procedures are sometimes introduced into practice without rigorous evaluation, and it is likely that at least some of them may be unjustified. Recently, Moseley et al. reported that in a PCT of arthroscopic surgery for osteoarthritis of the knee, the surgical intervention was no more effective than placebo operations (Moseley et al. 2002).

An in-depth analysis of this problem by ethicists from the National Institutes of Health suggests that surgery may indeed be associated with a strong placebo effect; this and other reasons discussed in this relevant and obviously very important paper lead the authors to believe that a placebo control may be required for a rigorous scientific evaluation of a surgical procedure – when the primary outcome is subjective (for example, pain or quality of life), especially as no other sufficiently rigorous study design poses less risk. The authors also emphasise the consequences of not conducting rigorous trials of surgery, in particular the exposure of patients to unjustified risk and the high costs incurred by the health system (over $3 billion in the USA alone, in the case discussed) without benefit to the patients. They believe that such PCT should be considered and conducted – following rigorous ethical assessment – before the procedure becomes standard (Horng and Miller 2002).

Recently, experts have also defined conditions that would permit the ethical use of placebo in osteoporosis: refusal of approved therapies by well-informed patients, substantial disagreement or lack of consensus about whether approved treatments are better than placebo or subjects are refractory to known effective agents (Brody et al. 2003). Those conditions may also apply to other situations where placebo use might be considered.

An international conference on placebo was held in Warsaw in April 2003 under the auspices of the Secretary General of the Council of Europe (http://surfer.iitd.pan.wroc.pl/events/Placebo.html).[1]

As already mentioned, Dr P. Zilgalvis, deputy head of the Bioethics Department of the Council of Europe, highlighted the "middle ground" approach (Zilgalvis 2004). The rationale for this policy on placebo has been formulated and proposed by Emanuel and Miller (Emanuel and Miller 2001). It appears that the acceptance of this approach would allow for the optimisation of the ethical and scientific conduct of clinical research.

1.
The programmes of other bioethical conferences held in Warsaw are available on the following websites:
http://surfer.iitd.pan.wroc.pl/events/integrity.html
http://surfer.iitd.pan.wroc.pl/events/misconduct.html
http://surfer.iitd.pan.wroc.pl/events/coi.html
http://surfer.iitd.pan.wroc.pl/events/patents.html

Bibliography

Barsky, A.J., Saintfort, R., Rogers, M.P. and Borus, J.F. "Non-specific medication side effects and the nocebo phenomenon", *Science & Engineering Ethics* 2004; 10 :133-134.

Box, J.E. "Placebos and MRC – and the consumer perspective", *Science & Engineering Ethics* 2004; 10 :95-101.

Brody, B.A., Dickey, N., Ellenberg, S.S., et al. 2003. "Is the use of placebo controls ethically permissible in clinical trials of agents intended to reduce fractures in osteoporosis?", *Journal of Bone and Mineral Research* 2003; 18 : 1105-9.

de la Fuente-Fernandez, R. and Stoessl, A.J. "The biochemical bases of the placebo effect", *Science & Engineering Ethics* 2004; 10 :143-150.

Ellenberg, S.S., Temple, R. et al. 2000. "Placebo-controlled trials and active-control trials in the evaluation of new treatments. II : Practical issues and specific cases", *Annals of Internal Medicine* 2000; 133 : 464-70.

Emanuel, E.J., Wendler, D. and Grady, Ch. 2000. "What makes clinical research ethical?", *Journal of the American Medical Association* 2000; 283 : 2701-11.

Emanuel, E.J. and Miller, F.G. 2001. "The ethics of placebo-controlled trials – a middle ground", *New England Journal of Medicine* 2001; 345 : 915-19.

Forster, H.P., Emanuel, E. and Grady, Ch. 2001. "The 2000 revision of the Declaration of Helsinki : a step forward or more confusion?", *The Lancet* 2001; 358 : 1449-53.

Gerold, R. "Ethics in research : striving for common ground within the cultural diversity of Europe", *Science & Engineering Ethics* 2004, 10 :5-8.

Horng, S. and Miller, F.G. 2002. "Is placebo surgery unethical?", *New England Journal of Medicine* 2002 ; 347: 137-9.

Hrobjartsson, A. and Gotzsche, P.C. 2001. "Is the placebo powerless? An analysis of clinical trials comparing placebo with no treatment", *New England Journal of Medicine* 2001; 344 : 1594-1602.

Huston, P. "What does the public think of placebo use? The Canadian experience", *Science & Engineering Ethics* 2004; 10:103-117.

Idanpaan-Heikkila, J. and Fluss, S. "The CIOMS view on the use of placebo in clinical trials", *Science & Engineering Ethics* 2004; 10:23-28.

Koski, G. and Nightingale, S. L. 2001. "Research involving human subjects in developing countries", *New England Journal of Medicine* 2001; 345: 136-8.

Michels, K.B. and Rothman, K.J. 2003. "Update on unethical use of placebos in randomised trials", *Bioethics* 2003; 17: 188-204.

Moseley, J.B., O'Malley, K., Petersen, N.J. et al. 2002. "A controlled trials of arthroscopic surgery for osteoarthritis of the knee", *New England Journal of Medicine* 2002; 347: 81-8.

Shapiro, H.T. and Meslin, E.M. 2001. "Ethical issues in the design and conduct of clinical trials in developing countries", *New England Journal of Medicine* 2001; 345: 139-42.

Singer, E.A. "The necessity and the value of placebo", *Science & Engineering Ethics* 2004; 10: 51-56.

Sugarman, J. "Using empirical data to inform the ethical evaluation of placebo-controlled trials", *Science & Engineering Ethics* 2004; 10: 29-35.

Temple, R.J. and Meyer, R. 2003. "Continued need for placebo in many cases, even when there is effective therapy", *Archives of Internal Medicine* 2003; 163: 371-3.

Temple, R., Ellenberg, S.S. et al. "Placebo-controlled trials and active-control trials in the evaluation of new treatments. I. Ethical and scientific issues", *Annals of Internal Medicine* 2000; 133: 455-63.

Tuffs, A., K. "Kurt Semm", *British Medical Journal* 2003; 327: 397.

Zilgalvis, P. "Placebo use in Council of Europe biomedical research instruments", *Science & Engineering Ethics* 2004; 10: 15-22.

Cancer clinical trials

by Maxime Seligmann

The European Directive 2001/20/EC, which is about to be incorporated into French law, states that

> [t]he accepted basis for the conduct of clinical trials in humans is founded in the protection of human rights and the dignity of the human being with regard to the application of biology and medicine" and that "[t]he clinical subject's protection is safeguarded through risk assessment based on the results of toxicological experiments prior to any clinical trial [and] screening by ethics committees.

Article 1 states that "[g]ood clinical practice is a set of internationally recognised ethical and scientific quality requirements [and that] compliance with this good practice provides assurance that the rights, safety and well-being of trial subjects are protected, and that the results of the clinical trials are credible".

Ethical considerations are particularly important in oncology, for two reasons: firstly, the patients are suffering from what is often a very serious (and feared) disease and, secondly, most of the methods used to treat it are highly cytotoxic* and have side-effects that are sometimes serious. The ethical problems are particularly acute in so-called phase I trials* which will be considered at the end of this chapter.

The various treatment methods

Most of the drugs currently used are cytotoxic agents, and the usual forms of chemotherapy combine, simultaneously or alternately, several such substances. A bone marrow transplant is increasingly indicated in many malignant blood disorders, and this requires prior "conditioning" by radiation and/or chemotherapy.

"Cytostatic" drugs are currently being developed on a large scale and probably represent the treatment of the future. Most of these drugs attack pharmacological targets that are essential for the proliferation of cancer cells. A very large number of drugs is involved, and they belong to two main categories: there are those which modify the biological behaviour of the

Cytotoxicity:
the quality of being poisonous to cells. This can be a chemical substance or an immune cell.

Phase I trial:
a clinical trial on normal volunteers, designed to determine the biological activities and range of toxicity or other safety factors of a given therapy.

Angiogenesis:
the formation of new
blood vessels.

tumour (for example, by blocking growth factors, hormones or signal pathways, or by inhibiting invasion), and those that alter the body's response (for example, immune-response modulators and angiogenesis* inhibitors).

Care and therapeutic research

The role of the doctor administering the treatment is to care for individual patients in what he or she considers the best possible way. Care is therefore a therapeutic activity which caters for the subjective needs of a particular person, whereas research is designed to take an objective approach to a biological individual. The former approach involves listening to someone in a way that takes account of personal concerns, whereas the latter is designed to update the impersonal laws that apply to human beings (French Advisory Committee on Ethics in the Life Sciences and Health Field [CCNE] 2003, Opinion No. 79). It is the elimination of everything that is purely subjective that allows knowledge of diseases to progress. "We must therefore accept the fact that there is a real divide between knowledge that takes account of subjective concerns and scientific knowledge" (Bachelard 1938).

Clinical trials are designed to answer scientific questions and check hypotheses, so as to expand our knowledge and improve treatment for future patients, and thus provide collective benefits. The divergence between the personal dimension of care and the collective benefits of therapeutic research affects the relationship between doctor and patient, which is built on trust. It means that there is an absolute duty to protect people who take part in clinical trials, and it raises ethical problems that were analysed very recently (Miller and Rosenstein 2003). What is unethical is not the fact of taking a collectivist approach to investigations, but to lose sight of the patient's own interests and treat him or her simply as a research subject.

The tendency of patients to confuse personalised medical care with participation in a clinical trial has been dubbed a "therapeutic misconception". Investigators must counteract this by informing patients as honestly as possible and in a way that is as easy to understand as possible. They themselves must guard against this confusion and be aware of the potential conflict of interests between scientific progress and the protection of research subjects. Integrity requires that they rule out any form of exploitation of a vulnerable subject. It is a moot point, moreover, whether the clinical trial investigator should not be separate from the doctor administering the treatment.

Assessment of the risk/benefit ratio

This is a key stage when a clinical trial protocol is devised and drafted. An analysis of the risks necessitates recognition of all the physical, psychological, social and financial implications. The pros and cons of participation in the trial and the constraints entailed must be carefully weighed up.

The benefit to the patient must be assessed in both quantitative and qualitative terms. A benefit cannot be assessed solely in terms of the number of months or years gained. The development of a new molecule or a new form of treatment that makes it possible to prolong the life of a patient suffering from an incurable disease may not bring about any improvement in his or her quality of life – an aspect we consider it essential to take into account. There is some substance in the view that "patients treated as part of a clinical trial do better than those receiving routine treatment" (Hoerni 1991), because they are more closely monitored. The concept of benefit must remain sufficiently flexible to allow biomedical research to oscillate between individual and collective concerns.

The practice of setting up a data and safety monitoring board comprising independent experts to oversee the trial should be encouraged for most trials.

Patient information and consent

European Directive 2001/20/EC states that the trial subject must have "given his written consent after being informed of the nature, significance, implications and risks of the clinical trial [and] may ... withdraw from the clinical trial at any time by revoking his informed consent". Consent is based on the information leaflet prepared by the sponsor, which must be as explicit and objective as possible, and on the information provided orally by the investigating doctor. The provision of information must take the form of a dialogue and exchange of views.

The information must make it clear that what is involved is a research project and that its benefits are uncertain, and the alternatives to participation in the trial must be put forward. The consent form must not be signed until several days after the information leaflet is handed to the patient, so that the latter may discuss the matter with his or her family and general practitioner, and until the investigator has answered any new or repeated questions.

In the case of paediatric oncology, the issue of minors deserves special consideration. In addition to the consent of the parents or legal representative, the investigator must expressly take account of the personal consent of under-age children who are able to make their wishes known and also take account of a refusal or the revocation of consent on their part. It is an ethical requirement to ascertain the wishes of the child.

The CCNE has issued an opinion (CCNE 1998, Opinion No. 58) on the "informed consent of and provision of information to persons taking part in treatment or research". Numerous articles on these issues have been published in connection with cancer clinical trials (Daugherty 1999; Tattersall 2001), and questionnaires designed to assess the quality of informed consent have been drafted (Joffe et al. 2001).

We consider it desirable that a European model information leaflet and consent form, covering the information that must be included, be drawn up to help investigators promote good practice.

Ethics committees

Ethics committees play a key role in protecting people taking part in research, for they provide a critical and ethical analysis of the therapeutic research envisaged.

Under European Directive 2001/20/EC, in preparing their reasoned opinions, such committees must consider, in particular: the relevance of the clinical trial and the trial design; whether the evaluation of the anticipated benefits and risks is satisfactory and whether the conclusions are justified; the suitability of the investigator and supporting staff; the adequacy and completeness of the written information to be given and the procedure to be followed for the purpose of obtaining informed consent; and the arrangements for the recruitment of subjects.

Ethics committees must be multidisciplinary and include persons with appropriate scientific expertise (doctors, scientists, pharmacologists in the case of phase I trials, and statisticians in the case of phase III* and IV* trials), but also nurses, legal experts, psychologists or psychiatrists and other representatives of civil society. To ensure independence and avoid the exercise of pressure, it is desirable that the bodies concerned be separate from the treatment centre where the principal investigator works.

Involvement of nursing teams and patients' associations

Involvement in a trial confers special responsibility not only on the doctor but also on the nursing team, all of whose members should be involved in the conduct of the trial and ensure, beforehand, that the patient has understood what is at stake. The involvement of patients' representatives is to be highly recommended when the protocol is devised and particularly when the information leaflet is drafted. It may also be desirable when the ethics committee examines the protocol.

As the trial ends

It is essential that arrangements be made for the long-term monitoring of subjects when they leave the trial, and for them to receive the treatment that has proved the most effective.

The findings of a clinical trial should be made public within a reasonable time. Even negative results should be published, given the need for openness in research. Failure to comply with these rules is unethical (Antes and Chalmers 2003). A European register of cancer clinical trials and their results is highly desirable.

Trials not involving medicinal products

The European directive concerns "clinical trials on medicinal products for human use" and does not therefore concern

Phase III trial: clinical trial using a large sample of patients, designed to compare the overall course of their disorder under the new treatment with its course untreated and treated with standard therapies previously used; studies are also done on the relative morbidities of the different treatments.

Phase IV trial: additional studies done after a drug has been approved for distribution or marketing, which could include examination of long-term effects, adverse effects, or specific aspects of a drug's action.

physio-pathological and cognitive oncological research. The provisions concerning information and consent should also apply to such research. There is justification for subjecting behavioural research to ethical scrutiny. Its consideration by ethics committees requires the presence of researchers in behavioural science. All research must be evaluated.

European multicentre trials

There are numerous financial, legal and administrative obstacles to large-scale trials carried out at European level by public institutions that are scientifically independent of the pharmaceutical industry (whose major objectives are to obtain product licences and market its products) (Cornu et al. 1999). The recent European directive, which is designed to simplify and harmonise statutory and administrative rules in the various member states, should make it possible to remove some of these obstacles. For instance, it obliges each member state to issue a single ethics committee opinion, and rightly states that repetitive trials should not be carried out within the Community.

Until very recently, the European Commission had serious misgivings about financing clinical trials. Private and quasi-public bodies, such as the European Organisation for Research and Treatment of Cancer in the cancer field, have endeavoured to remedy this situation.

Phase I trials

The CCNE recently published an opinion on the ethical issues raised by phase I cancer trials (CCNE 2002, Opinion No. 73).

Phase I trials are defined as the first trials carried out on human beings after animal and *in vitro* experiments. They are an essential stage in the use of any new molecule, and their main purpose is not to look for a therapeutic effect, but to assess toxicity. They are therefore designed to reveal any side-effects, their duration, whether or not they are reversible and their relationship with pharmacokinetic data. The data obtained are needed in order to carry out the initial studies on the effectiveness of the drug.

As anti-cancer molecules are generally highly cytotoxic, phase I trials cannot be carried out on healthy volunteers, and are conducted on cancer patients for whom no further treatment is available. In phase I trials, it is compulsory to administer gradually increasing doses, on the grounds that the highest dose tolerated is the one which is the most likely to be effective. This widely accepted hypothesis is by no means always confirmed. The traditional method, whereby at least three patients receive the molecule at each dose, is now rarely used.

Other systems for increasing the doses, using new statistical models and new pharmacokinetic methods, have been proposed. The purpose of these changes is to determine the toxic dose more quickly, avoid an excessive risk of toxicity and limit the number of patients to whom a dose that is very low and therefore, a priori, completely ineffective is administered. Despite progress, however, it is difficult to achieve these three objectives simultaneously (Eisenhauer et al. 2000).

There is little point in determining the maximum dose tolerated in the case of the various cytostatic agents, which can be most effectively administered by means of sub-acute exposure or chronic exposure, whether continuous or weakly discontinuous. Instead, in the case of these molecules, the objective is to determine the biologically effective dose, which is usually very different from the toxic dose. The recent guidelines of the European Agency for the Evaluation of Medicinal Products (Emea 2003), which cover both cytotoxic and cytostatic agents, continue, however, to state the need to determine the maximum tolerated dose, even though this requirement has been criticised by certain authors (Degos 2000).

Methodological and pharmacological progress should make it possible for tolerance and effectiveness to coexist. At present, however, the objective of these preliminary but necessary phase I trials is to assess tolerance to the new molecule without directly determining whether there is a therapeutic benefit to the patient taking part. This shows how heavily weighted the risk/benefit ratio is in favour of the risk. Such trials are therefore at variance with the Declaration of Helsinki, although the person drafting the protocol must indicate that he or she has complied with the declaration, which states that "[i]n medical research on human subjects, considerations related to the well-being of the human subject should take precedence over the interests of science and society" (World Medical Association Declaration of Helsinki 2000).

Side-effects are frequent and sometimes serious, whereas the clinical benefit is very small. Specialists differ over its frequency and scale. Meta-analyses show that fewer than 6% of patients derive a (minor or more substantial) benefit, whereas 0.5% of them die because of the product's toxicity (Horng et al. 2002). The benefit is, of course, more substantial when the trial combines a new molecule and a previously recognised medicinal product.

Pharmacological progress will make it possible to identify patients who are unlikely to benefit from the new molecule or, conversely, those whose tumours will be sensitive to the substance being tested. Moreover, it is desirable that, as far as possible, patients taking part in phase I trials should, secondarily, be able to benefit as soon as possible from a phase II trial designed to determine the effectiveness of the substance.

The choice of patients is an ethical issue of key importance. Such trials are normally offered to patients for whom there is no alternative treatment. It is, however, highly desirable that terminally-ill patients should not take part, for both scientific and ethical reasons. In terminally-ill patients, clinical and biological tolerance and pharmacokinetic data on cytotoxic molecules may be different from those in a less severely affected patient, and this calls into question the actual scientific validity of the proposed protocol. Besides, the inclusion of particularly vulnerable people, who are often prepared to submit to any phase I trial without having properly understood its purpose and scope, raises a real ethical problem.

The eligibility criteria of the European Agency for the Evaluation of Medicinal Products (Emea 2003) require that the survival probability exceed eight to twelve weeks. It will probably become possible to select patients likely to benefit from a new molecule by identifying targets and by means of pharmacogenetic studies. This might mean that it would not be suggested that a patient be included in a phase I trial until a tumour sample has been taken in order for these parameters to be studied.

It is particularly important to inform patients to whom such trials are suggested and obtain their informed consent, and these procedures often cause many problems. There is a wealth of literature on the subject in English: the main references are to be found in the article by Horng et al. (2002). There is a major conflict of interests between the need to find cancer patients in order to explore, in phase I, tolerance to new molecules and the duty to care for individual patients, and this presents the doctor with a genuine moral dilemma (Miller 2000).

The doctor must indicate, in the information leaflet and at the interview with the patient, that the main objective of the trial is to determine tolerance to a new substance, and must avoid the word "treatment". He or she must specify the nature of the toxic symptoms that will be looked for and their possible effects on the patient's quality of life. The doctor must refer to a modest hope that there will be benefits, without concealing the uncertainty surrounding this. He or she must also discuss alternatives, such as continuing a standard form of treatment, which will most probably be ineffective, stopping all treatment, and palliative care.

It is necessary to avoid any ambiguity that allows the problem to be played down, deliberately or otherwise. The concept of collective utility and of benefits to other patients must be referred to. Each patient must understand that any new treatment is invariably based on such trials. The patient's main motivation, however, is not generally altruism, but the hope of a new form of treatment. The main problem is to avoid dashing any hopes, without raising them unjustifiably. Such rigorous openness must, however, take account of the patient's capacity for discernment. In its Opinion No. 58, the CCNE stressed that "the duty to inform does not imply the right to do so bluntly or abruptly".

It is always necessary to consider whether patients' consent is genuinely independent, as the independence of their judgment is undermined by the vulnerable situation in which they find themselves. The main ethical problem raised by phase I trials is therefore that of ensuring that the decision to include a patient is the result of a genuine exchange between doctor and patient in a relationship based on mutual trust. The factors involved in the patient's decision have been analysed in a major centre in the United States (Gordon and Daugherty 2001). The conditions under which trials are proposed and consent is obtained naturally raise particularly serious issues in the case of phase I paediatric oncology trials.

As regards European regulations, it is desirable that, in the case of molecules that have already been tested and used abroad, account be taken of prior phase I results so as not to repeat trials unnecessarily in patients, who would thus be subjected to administrative rather than scientific constraints.

> The duty of solidarity must not be used as an argument for undermining the rights of the individual. Society as a whole must be aware that research needs may sometimes cause the interests of the community to take precedence. This awareness is never, however, a reason for ignoring the important requirement that full respect be shown for the individual who, by virtue of his or her disease, may help humanity (CCNE 2002, Opinion No. 73).

Bibliography

Antes, G. and Chalmers, I. 2003. "Under-reporting of clinical trials is unethical", *the Lancet*, 361; 9362 : 978-9.

Bachelard, G. 1938, *La Formation de l'esprit scientifique*, Paris, Vrin, p. 239.

CCNE 1998, Opinion No. 58, *Consentement éclairé et information des personnes qui se prêtent à des actes de soin ou de recherche*, available on the Web: http ://www.ccne-ethique.fr

CCNE 2002, Opinion No. 73, *Les essais de phase I en cancérologie*, available on the Web: http ://www.ccne-ethique.fr

CCNE 2003, Opinion No. 79, *Transposition en droit français de la directive européenne relative aux essais cliniques de médicaments : un nouveau cadre éthique pour la recherche sur l'homme*, available on the Web: http ://www.ccne-ethique.fr

Cornu, C., Cano, A., Pornel, B., Melis, G.B. and Boissel, J.P., on behalf of the EUTERP Pilot Study Group, 1999. "Could institutional clinical trials exist in Europe?", *the Lancet*, 353 ; 9146 : 63-4.

Daugherty, C.K. 1999. "Impact of therapeutic research on informed consent and the ethics of clinical trials : a medical oncology perspective", *Journal of Clinical Oncology*, 17 :1601-17.

Declaration of Helsinki, 2000, available on the Web: http ://www.wma.net

Degos, L. 2000. "Phase I trials in cancer treatment. The maximum tolerated dose : a barbarian guideline", *Hematology Journal*, 1; 4 : 219.

Directive 2001/20/EC of the European Parliament and of the Council of 4 April 2001 on the approximation of the laws, regulations and administrative provisions of the Member States relating to the implementation of good clinical practice in the conduct of clinical trials on medicinal products in human use, O.J. L 121/34 of 1.5.2001, pp. 34-44, available on the Web: http ://europa.eu.int/eur-lex

Eisenhauer, E.A., O'Dwyer, P.J., Christian, M. and Humphrey, J.S. 2000. "Phase I clinical trial design in cancer drug development", *Journal of Clinical Oncology*, 18; 3 : 684-92.

Emea, 2003, note for guidance on evaluation of anticancer medicinal products in man, available on the Web: http ://www.emea. eu.int/ pdfs/human/ewp/020595en.pdf

Gordon, E.J. and Daugherty, C.K. 2001. "Referral and decision making among advanced cancer patients participating in phase I trials at a single institution", *Journal of Clinical Ethics*, 12; 1 : 31-8.

Hoerni, B. 1991, *L'Autonomie en médecine. Nouvelles relations entre les personnes malades et les personnes soignantes*, Paris, Payot, p. 118.

Horng, S., Emanuel, E.J., Wilfond, B., Rackoff, J., Martz, K. and Grady, C. 2002. *Descriptions of benefits and risks in consent forms for phase I oncology trials*, N 2134-40.

Joffe, S., Cook, E.F., Cleary, P.D., Clark, J.W. and Weeks, J.C. 2001. "Quality of informed consent in cancer clinical trials : a cross-sectional survey", *the Lancet*, 358; 9295 : 1772-7.

Miller, M. 2000. "Phase I cancer trials : a collusion of misunderstanding", *Hastings Center Report* 30; 4 : 34-42.

Miller, F.G. and Rosenstein, D.L. 2003. "The therapeutic orientation to clinical trials", *New England Journal of Medicine*, 348; 14 : 1383-8.

Tattersall, M.H.N. 2001. "Examining informed consent to cancer clinical trials", *the Lancet*, 358; 9295 : 1742-3.

Some ethical considerations in industry-sponsored clinical trials

by Tom Gallacher and Sreeharan[1]

> "Ethics and Science need to shake hands"
> *Richard Clarke Cabot (1868-1939)*

This statement becomes truer with the passage of time. The pharmaceutical industry perspective on biothics derives from the constant need to integrate ethical principles with rapidly evolving scientific and medical advances and hence new investigational paradigms.

The primary societal responsibility of the pharmaceutical industry is to discover and develop new drugs, vaccines and diagnostics. Biomedical research is an important element of global strategies to improve health and healthcare. However, taking account of patient needs in the context of research ethics is also critical in determining industry research and development (R&D) strategies. The time taken from initial discovery to product launch can be several years and the estimated average cost of development can be of the order of 800 million euros or more.

As well as being a long, costly process, pharmaceutical R&D is complex and risky. For each compound that is eventually licensed for sale, several thousand will have been identified and discarded along the way. In the course of a development programme there are many check points, both in the pre-clinical stage and once the potential product has progressed to evaluation in humans. Throughout this lengthy process, assessments of the risk/benefit profile are made at frequent intervals, with the possibility of the project being terminated at any stage if an unfavourable ratio is detected.

The development programme is conducted within a very demanding nationally and internationally regulated framework. The need for product safety to be paramount has produced a regulatory environment for pharmaceuticals that is one of the most demanding of any industry. In addition to the external

1.
The views expressed are the personal views of the authors and do not necessarily represent the policy or position of Glaxo-SmithKline nor the pharmaceutical industry in general.

scrutiny of government regulatory agencies during develop-ment and following submission of an application for a product licence, most research-based pharmaceutical companies have some form of safety board charged with the overview of human safety issues associated with the development of potential new products. Typically this body will be chaired by the chief medical officer and its membership will comprise senior physicians with a facility to involve, either routinely or as needed, independent, external experts.

Innovative new medicines, by the very nature of their targets being novel, may not deliver their full potential to healthcare and patient needs at the time of introduction of the product. This is one important reason why pharmaceutical companies continue to sponsor clinical trials after products are initially launched. In many instances, additional uses emerge from this post-marketing stage of research; for example, the role of ACE inhibitors in the treatment of heart failure, following their initial introduction as anti-hypertensives. Hence industry sponsorship of clinical research can often continue over the majority of the product life cycle.

In order to improve this prospect, the industry exercises a pre-cautionary approach when conducting clinical trials, with a high level of internal scrutiny and quality control in addition to the scrutiny of external regulatory agencies. This paper pre-sents a perspective on how industry approaches some of the ethical issues associated with clinical trials which it sponsors. In developed countries, industry-sponsored clinical trials account for the majority of all clinical trial activity.

The ethical justification of biomedical research involving human subjects is the prospect of adding to scientific knowledge which will enhance healthcare. Additionally, to be ethical, the investigation must be designed in a scientifically robust manner and conducted in a way that ensures respect for the dignity and safety of participants as judged morally acceptable, relative to international norms and by the locality/communities in which it will be conducted.

There are a number of international instruments on the ethics of medical research of which the Nuremberg Code, the Belmont

Report, the UN Declaration of Human Rights and the World Medical Association's Declaration of Helsinki in particular contribute to the key principles that are widely accepted by medical researchers including the pharmaceutical industry as research sponsors. The key principles are complemented by other declarations and several operational-level guidelines. From an industry perspective, the ICH guidelines on Good Clinical Practice (GCP) – together with relevant sections of the US Code of Federal Regulations and other national or regional regulations – have particular relevance, as they set standards with which industry-sponsored research has to comply.

Recognition of the need to respect the dignity of trial participants and protect their safety are fundamental objectives for all clinical-research sponsors (including industry) and investigators.

Ethics-related areas or issues that are reviewed in this paper are :

- the role of ethics committees in overseeing clinical research;

- the informed consent process;

- conflict of interest and payments to investigators and trial participants;

- research in populations/communities with limited resources.

Role of ethics committees

The mechanism that is universally used to assess adherence to the fundamental ethical norms is that of independent ethics committees (in some countries referred to as independent review boards). Ethics committees have independent decision-making authority and the responsibility to act in the interests of research participants. Proposed clinical trial designs and implementation plans (in the form of protocols) must be submitted to an ethics committee for review before clinical trials can commence. The ethics committee has the right to endorse a proposed clinical trial, or require changes, or disapprove it. Confirmation of their favourable opinion is a required step before clinical trials can commence.

There are well-defined considerations against which the ethics committees evaluate the protocol and operational aspects of the trial. Two key elements are:

• the information to be provided to the potential participants; and

• whether there could be inducements that might unduly influence either the investigators or the potential participants.

One of the obligations of ethics committees is to assess the adequacy of the information to be provided during the informed consent process. This needs to be done on a case-by-case basis, relative to the nature of the drug and/or test procedures involved. There is no "one size fits all" set of information for informed consent statements, although some of the information will be common to any trial situation.

A number of international guidelines as well as national laws/regulations provide details of what should be explained in the information to be provided to potential participants. Among the essential elements are:

• clear communication that the trial involves research (including any aspects that are experimental);

• the purpose of the trial;

• whether and to what extent participants will be assigned to one of several treatment regimens;

• reasonable foreseeable risks or inconveniences;

• reasonably expected benefits – if none, participants must be made aware of this;

• alternative treatments that may be available;

• any anticipated reimbursement of expenses or payments for participation;

• that participation is voluntary and refusal or withdrawal from the study will not affect their future healthcare.

More extensive lists of elements to be considered for inclusion can be found for example in ICH GCP, the US Code of Federal Regulations and the CIOMS International Guidelines for Biomedical Research Involving Human Subjects.

Any proposed deviation from full disclosure of study procedures must be explicitly identified to the ethics committee and be endorsed; for example, the purpose of tests performed to monitor participants' compliance with the protocol. Similarly, if the normal requirement that participants provide personal consent might not be possible, as in emergency situations such as stroke or head trauma, the proposed alternative process must be brought to the ethics committee's attention so that they can consider whether it is justified and advise accordingly.

Given the critical role of ethics committees in the overview of clinical trials, the pharmaceutical industry recognises the need for sufficient, well-trained, well-resourced committees capable of thorough protocol review. Whenever possible structured, sustainable initiatives to develop increased review capacity need to be encouraged.

Informed consent process

The original concept of consent to medical care related to a doctor's obligations to obtain the patient's permission before intervening medically. "Consent" has evolved to "informed consent", reflecting the increasing emphasis on patients being informed about risks. In the context of medical research, informed consent takes on additional meaning and complexity because the individual is invited to participate in a research project which, to varying degrees, is experimental. In this setting, it is considered necessary to ensure that potential participants are adequately informed of potential risks and benefits of the trial.

In their role as clinical-trial sponsors, pharmaceutical companies accept that they have a duty to ensure that investigators establish a process for obtaining individual, informed consent from potential participants in a manner that is appropriate to the trial circumstances, for example the medical condition being treated, the patient population and any relevant cultural factors. To meet these requirements the person obtaining consent has to be sufficiently knowledgeable about the trial to be able to deal with participants' questions. To ensure that this will be the case, industry sponsors will address this area within the overall training/briefing of investigational site staff that is undertaken prior to the trial starting. This will extend to knowledge about the trial protocol, product information and procedural issues, including adherence to Good Clinical Practice (GCP) requirements and safety monitoring and reporting obligations.

Consent should be sought only once it is clear that the potential participant has received and understood the information relating to the trial (as endorsed by the ethics committee), and has considered the options. There must be no undue influence. Both before and during the trial there should not be any obstacles that could restrict the opportunity for participants to have questions answered/clarified.

Individual, informed consent must be documented and, unless there are circumstances that have been deemed justifiable by the ethics committee, documentation will be by way of a consent form personally signed by each participant. Circumstances in which it might be justified to document consent in an alternative manner could be where a participant is unable to read. In this scenario an impartial witness should be present during the consent process in order to sign the consent form, thereby attesting that the trial-related information was accurately explained, that it was apparently understood by the participant (or their legally acceptable representative) and that the participant (or legally acceptable representative) gave their consent voluntarily.

In the case of trials involving children who by law are not able to "consent", agreement to participate must be by a parent/guardian/legal representative of the child. Also, when age and mental comprehension allows, the assent of the child should be obtained following an explanation of the trial in terms relevant to their comprehension.

Conflict of interest and payments to investigators and trial participants

The issue of payments both to investigators and trial participants continues to be controversial. In the case of payments to research participants, views range from "it is never ethical" to "ethically acceptable provided the amounts do not constitute inducement". Industry is increasingly sensitive to the possibility that ethics committees and others may consider any payments as potential inducements. This is therefore an area where the ethics committee is in the best position to make an objective assessment as to whether the level of proposed payment is appropriate and unlikely to induce participation for economic benefit. These are not always easy judgments to make and the need for knowledge of local circumstances is another reason why ethics committees are the appropriate forum to assess this.

With regard to financial and other influences on those involved in clinical trials, different considerations apply in the case of investigators/researchers – if they receive payment – and in the case of participants.

Payments to investigators/researchers

When sponsor organisations agree with investigators that they will participate in the conduct of a clinical trial, it is entirely reasonable that they should be reimbursed for the services that they contribute in conducting the study, including any reasonable personal expenses such as travel, with the proviso that any payments do not create a personal conflict of interest and that they satisfy the criterion of being in accordance with "fair market value".

Payments to researchers and/or their institutions must be formally agreed and documented. There should be a clear understanding as to the purpose of the overall payment – salaries of research workers, technicians and nurses, administrative support, equipment purchase or hire, institutional overhead costs, attendance of research staff at scientific meetings for study related activities.

Payments should be commensurate with the time and effort which will be required, which means application of the concept of "fair market value". In the context of pre-enrolment assessment of potential participants, payments should not exceed the cost of assessing participant's suitability plus any associated expenses.

As already identified, this is an important aspect of the information to be assessed by the ethics committee.

Payments to the study participants

There is a long tradition that both patients and healthy volunteers for non-therapeutic studies are motivated by a sense of altruism and a desire to help society at large by contributing to the advance of knowledge. Despite this, it has been the practice to at least reimburse participants' "out-of-pocket" expenses and sometimes to pay for their participation. Different situations justify different ways of dealing with the issue of payments to participants. The assessment of what is appropriate in a given situation is again within the remit of the ethics committee, which must decide if the reimbursement and/or payment offered constitutes an undue inducement.

In the case, for example, of pharmacology studies that will offer no therapeutic benefit to participants, and which will sometimes involve intensive monitoring, it is accepted as reasonable that participants should be paid for the inconvenience and possible discomfort in addition to the reimbursement of any expenses incurred, such as transport costs. These reimbursements are designed to eliminate barriers to participation by returning the economic circumstances of participants to what they would otherwise have been. To avoid potential participants volunteering against their better judgment, payment should not be based on perceived exposure to risk.

In the case of studies that involve patients who will have an opportunity for therapeutic benefit, there are various potentially motivating factors that need to be considered in order to minimise the risk of undue inducement. For example, the requirement to consent should only be discussed relative to a specific trial protocol – not as consent to participate in research in a generic sense, followed by enrolment into a specific study that might provide access to a potentially beneficial treatment. For some patients it is undoubtedly the case that the opportunity to participate in a clinical trial will be seen as resulting in more detailed medical monitoring and supervision. In presenting the opportunity to participate, it is important that the prospect of improved care is not offered in a way that constitutes an inducement.

It is also important to minimise the effects of both the patient's loyalty to their doctor or the "power dynamics" of the doctor/patient relationship as factors that could unduly influence patients to decide to participate. There are two important assurances that must be communicated to potential participants:

- the patient is entirely free to decline to participate – this is their right;

- a refusal to participate will be accepted, and medical care will then continue in exactly the same way as though participation in a trial had not been considered.

A related issue is that of advertisements for participants. In countries where this is allowed, it is essential that applicable stipulations are followed. The advertisements should be constructed as an information item, not as an encouragement to participate. All proposed advertisements (print, audio or audio-visual) must also be submitted to the ethics committee for review prior to use, which provides a mechanism for independent scrutiny of this aspect.

Both with respect to payments to investigators/institutions and to trial participants, the role of the ethics committee is critical in assessing whether or not either form of payments might constitute inducement or conflict of interest. In doing so it might be necessary to also consider (in the case of potential participants) their likely medical and economic status together with any relevant cultural issues. The multifactorial nature of such an assessment for a particular trial means that it is not possible to have a "one size fits all" set of payment levels for all trials – even at a national level.

Clinical trials in populations with limited resources

In some instances it is claimed that pharmaceutical companies (and other sponsor organisations) have exploited the vulnerability of populations in resource-poor, developing countries in order to expedite global development programmes for products that will not become available to the populations of the resource-poor countries, or carry out research that would not be sanctioned in the country where the sponsor organisation is based.

There is now a high level of awareness that proposals to conduct clinical research in such settings have to take several key factors into account from the earliest consideration, for example:

- Is the research programme compatible with and responsive to the health needs and priorities of the population in which it is proposed to be conducted?

- Will the outcome (product or information) become available for the benefit of the host community/country?

- Is the research conducted to internationally accepted ethical standards?

- Will the research contribute to local infrastructure or capability development?

When sponsors consider proposing that clinical research projects should be conducted in areas with limited healthcare resources, it is therefore necessary to investigate the extent to which the project aligns with local priorities, and the prospect of meeting the factors such as those identified above. Local health priorities will be centred on diseases of the developing world and there is increased recognition within industry that ways must be found to make full use of existing products as well as develop relevant new products for these diseases. As part of its social contract industry is increasingly involved, either on its own or in partnership with the public sector, with R&D into diseases of the developing world.

Proper assessment of the issues will often be difficult and time-consuming, as it must involve relevant local stakeholders. According to specific circumstances, this might need to include relevant government agencies/departments, local health authorities, representatives of the communities that might be involved, potential investigators, representatives of concerned scientific and ethics groups, and other non-governmental agencies. Experience has shown that the most effective way to achieve this is on a partnership basis, through engagement of the sponsor(s) with the relevant parties.

Consultation needs to cover all aspects of the relevance of the research, its feasibility relative to the local healthcare infrastructure and any interventions needed to make it viable, confirmation that local ethics review capacity exists, what the healthcare provision for research participants will be after the trial and which parties will be responsible for these healthcare provisions. The general standards of conduct that should apply in such projects will be the same as for clinical trials in developed countries. This consultation process might identify opportunities for capacity building associated with a clinical trial and, as in the case of the trial itself, any specific capacity-building activities that are to be pursued should be agreed with the relevant local stakeholders.

The longer-term availability of the product (or information generated) is also an important aspect of the early stage consultation. This is a potentially complex issue and can only be determined on a case-by-case basis. It needs to take account of the projected costs of the product relative to the healthcare system and its capacity to provide the product. At this stage, cost projections may only be a provisional estimate due, for example, to the early stage of development or the unknown outcome of planned scale-up of manufacturing processes.

When, following initial consultation and feasibility assessments, it is proposed to progress further, all the information from these preliminary consultations should be made available to the local ethics committee considering the proposed project so that all relevant factors are considered. If it is clear that the outcome of a successful programme is not going to be reasonably available to benefit the host country once the overall development programme is completed, it is unlikely that there is an ethical justification to conduct the research in such locations. If it is considered that in exceptional circumstances such a proposal can be justified, its acceptability must be endorsed by the local ethics committee.

It will be clear from what is involved in assessing the feasibility of clinical research projects in developing countries that the best prospect for success with such projects is for them to be conducted on a partnership basis.

Women in biomedical research

by Outi Leena L. Hovatta

Women as study subjects

Requirements for clinical studies

The ethical principles regarding all clinical studies are the same for female and male study subjects. They are clear from the Declaration of Helsinki. Human dignity has to be respected in all medical research. The benefit to the individual is always more important than that to society or science. All the possible risks of a research project have to be avoided as much as possible. A study subject can only be exposed to procedures from which the expected benefit is greater than the possible risk or harm. The study subject has to be given information regarding the aim of the research, how the procedure is to be carried out, and the benefits and possible risks and harm associated with a procedure.

The information has to be given in such a way that the study subject understands it and is capable of giving her or his consent, being aware of the benefits and risks. Documented consent from the study subject is required. An exception can only be made for authorising the participation of a person not able to consent if the results of the research have the potential to produce real and direct benefit for her or his health.[1] The study subject always has the right to withdraw her/his consent. Information regarding this right has to be given to her/him.

Women and pregnancy, and the effects of treatment on the foetus

Clinical studies involving women as study subjects are very different from those involving only men, because women of fertile age may be pregnant or could become so.

Pregnancy is a very special period as regards medical treatment and research. The possible effects of pharmaceutical and other substances on the foetus are a major concern. Such concerns became real after the disasters that happened with thalidomide and diethylstilbestrol, both of which were used as medication

1.
The convention allows for research without the potential to produce results of direct benefit for the person concerned only when the research meets the strict protective conditions set out therein and only when it entails minimal risk and minimal burden for the individual concerned.

for pregnant women. Animal experiments had not indicated that these substances cause abnormalities during early human development, and large numbers of malformed infants were born before the causes were detected. That individuals exposed to diethylstilbestrol during foetal life can develop cancers in later life was, indeed, a most unpleasant complication of such treatment.

These unfortunate events have resulted in strict national and international regulations on how substances have to be studied before they can be used in clinical research and treatment, and there are particular requirements for tests for the possible harmful effects of substances on the foetus. In the European Union, these requirements are presented on the EU website and have been harmonised with corresponding requirements in the USA and in Japan. There are detailed guidelines and requirements regarding how to test the safety of a substance before any clinical trials are proposed. There are also special tests regarding reproductive toxicity that have to be performed.

The methods include certain numbers of tests on pregnant animals, and tests concerning the normality of the offspring. Mouse embryos can be cultured *in vitro*, and the effects of substances studied in such circumstances. There are also tests that can be performed in cell cultures.

In spite of the available tests, the possibility still remains that cell culture and animal tests will not reveal the possible effects on a human embryo and foetus. There are often differences between species, and in some respects human development is different from that in other species, including non-human primates. Human embryos cannot be used in toxicological tests, and mouse embryos may react differently. Hence, the concern remains that a pharmacological or chemical substance may cause disturbances in human embryonic or foetal development in spite of normal results in animal tests.

In addition to international requirements, each country has national laws that have to be adhered to before the statutory authority can give permission to carry out a clinical trial. In all countries in the European Union and in most other countries, ethics committee approval is needed before any study involving human individuals can be initiated. International biomedical journals do not publish articles based on studies that do not have ethics approval. These requirements have to be met in all studies involving human individuals, both women and men.

Concern regarding the possible effects on the foetus has resulted in extreme care in giving any medicines during pregnancy. There are lists of medicines that are known to be safe, and lists of medicines that are known to cause foetal abnormalities. These lists are available to medically and pharmacologically trained personnel, and also for other people. However, the list of medicines that have not been studied at all during pregnancy is much longer than the two other lists.

The possible influences of the vast majority of pharmacological substances on foetal development are unknown. Knowledge only accumulates from situations in which a woman has taken the medicine not knowing that she is pregnant, or in which the nature of her disease has demanded treatment in spite of the pregnancy. The general recommendation is to avoid medicines during pregnancy.

This strategy appears to be wise. On the other hand, it means that women often do not receive optimal treatment of their diseases during pregnancy. In many cases this is harmful to the woman's health, and may cause unnecessary suffering. New methods to study the effects of pharmaceuticals and chemicals on embryonic development in humans are definitely needed.

There is one new option to study the effects of various substances on early human development. Culture of human embryonic stem cells has been possible since 1998, when Thomson et al. reported their first permanent human embryonic stem cell lines (Thomson et al. 1998). Such cell lines have their origin in certain cells of human embryos five to seven days after *in vitro* fertilisation, this being the most effective treatment for involuntary childlessness. Now and then there are embryos that cannot be used in infertility treatment, usually because of poor quality. There are also embryos that have been frozen after the family has achieved the desired number of children after earlier treatments. These embryos are normally discarded.

Supernumerary embryos: surplus embryos as a result of *in vitro* fertilisation.

Blastocyst: (embryonic stage) : a mammalian embryo at the stage where it is implanted in the wall of the womb.

Cells can be taken from the inner cell mass of such embryos before they are discarded. One fifth of these supernumerary embryos* reach the so-called blastocyst* stage when cultured up to the age of five to seven days. From 10-15% of such blastocysts, the cells begin to grow in certain culture conditions, and they form the so-called cell lines. The difference between these early non-differentiated embryonic cells and more specialised cells is that the embryonic cells can divide and form new similar cells indefinitely, while specialised cells can only survive for certain numbers of generations. This means that one embryonic stem cell line can in optimal conditions form enormous numbers of cells.

Because these cells are the progenitors of all the cell types and tissues in the body, there are expectations that they can be used in cell transplantation in order to cure many severe diseases. However, being early human embryonic cells they might also be used in testing the effects of various substances on early human embryonic development. Such tests could be carried out in cell cultures. The cultures might be much more reliable than animal tests, and large numbers of animal experiments might become obsolete. But above all, the use of human embryonic stem cells would, in this respect, greatly benefit women's health.

There are, however, differing opinions as regards the ethical acceptability of the use of human embryonic stem cells. In this field, there is real plurality in Europe. The European Group on Ethics (EGE)[1] for the European Parliament and European Commission has prepared a thorough report regarding the ethics of working with human embryonic stem cells.

1. The EGE is an independent, pluralist and multidisciplinary body which advises the European Commission on ethical aspects of science and new technologies in connection with the preparation and implementation of Community legislation or policies.

Pregnant women as study subjects

Pregnancy can be regarded as a vulnerable situation for both the foetus and the mother. Hence, more protection is necessary for pregnant women than non-pregnant ones. This is also reflected in laws regulating medical research. In Finland, for instance, the law regarding medical research (9.4.1999/488, www.finlex.fi) regulates studies in which pregnant women can be subjects of medical research, and at the same time it regulates the conditions under which children or mentally impaired individuals can be recruited to become study subjects. A pregnant or breast-feeding woman can be a study subject only if the same scientific results cannot be obtained with non-pregnant subjects. The research project has to benefit the health of the woman, or the infant to be born, or it has to benefit the health of her relative(s) or other pregnant or breast-feeding women, or foetuses or newborn infants.

It is extremely important for women that as much as possible is known about the physiological and psychological events during pregnancy. There are very many changes in the human body during pregnancy, and they can only be measured during pregnancy. Pregnancy-related diseases can, of course, be studied only in pregnant women.

Numerous standard laboratory tests give different values during normal pregnancy, and the values change during the course of pregnancy. Many of them have been systematically studied by taking samples from pregnant women. However, there are some parameters that have still not been properly characterised, and they can cause serious concern among pregnant women. This is particularly the case with some rare diseases that may occur during pregnancy. This is very unfortunate and can result in poor treatment or unnecessary treatment of pregnant women. Obstetric disorders, related to pregnancy itself, are well-known and studied, and they have not been causing major concern.

Differences between female and male study subjects

The fear of embryonic effects also has other consequences on women's treatment. Differences between men and women in the effects and metabolism of many pharmaceuticals are poorly known, because it has been easier and safer to carry out clinical trials on men only. Women aged between 15 and 50 have often not been included because of the possibility of pregnancy, which is difficult to exclude completely. In addition, contraceptive pills, for instance, may influence the metabolism of other medicines. Women who are younger than 15 are children, and they cannot represent adult women as study subjects. Women over 50 are post-menopausal and their bodies again behave differently in many respects. The present information regarding the way in which many pharmaceuticals behave in the female body is based on extrapolation of what has happened in male study subjects. However, there is existing information that there are differences.

The hormonal variations during the menstrual cycle also make women different from men as study subjects. Different concentrations of many hormones, such as oestrogen and progesterone, influence many other common test parameters. There is a vast amount of literature in which this fact has not been taken into account at all. In recent years most clinical scientists have been aware of this fact, and they have been taking samples in both the follicular phase (time before ovulation) and the luteal phase (time after ovulation) of the cycle. When interpreting research results one has to be careful in this respect.

The effects of contraceptive pills on the female body, and the co-effects with other medicines, have been studied relatively thoroughly because they are regulated pharmaceutical substances. The same holds true as regards hormonal replacement therapy in post-menopausal women. For the time being there remains much to be known, but there are large clinical and epidemiological studies going on. These are of extreme importance for women's health.

Pregnancy also offers some health benefits to women. The probability of certain cancers arising is somewhat lower among women who have been pregnant. Some protection is obtained, for instance, against cancers of the ovary, uterus and the breast. Epidemiologists, in their large studies, take into account pregnancies as a contributing factor in various diseases.

It is not possible to study women's health and diseases, and to develop optimal treatment, without having women as study subjects. Pregnancies are an integral part of women's life and health, and it is of the uttermost importance to study pregnancy-related diseases with pregnant women as study subjects.

Women as scientists

Women are contributing to biomedical research not only as study subjects. Their role as researchers is, of course, most

important. It is a natural thought that 50% of biomedical scientists would be women, if half the population are female. This
is not yet the reality in many European countries, but efforts
are being made to achieve it. Getting the female half of the
population actively involved in research has been mentioned
in European science reports (EU 2002 ; EU 2003).

Having female scientists in leading positions is important
when thinking of the problems that still exist in women's
health, and equal treatment. Female scientists are more likely
to pay attention to these particular problems.

The proportion of female university students has been growing
everywhere. In many countries, including Finland and Sweden,
there are now more female than male graduates from the universities in the fields of medicine and biology. Among students
and graduates in technical universities the proportion is still
low, but it is growing. This could mean that in the future the
proportions among leading scientists may be equal. Active fol-
low-up and intervention, however, are necessary because there
are other factors apart from the numbers of trained persons
who contribute to positions in the research community.

There are statistics regarding the proportion of female professors
in universities in different countries. In the countries of the
European Union, the proportion of women is still much lower
than that of men. There have been wide discussions about this
in Sweden, with attempts to actively increase the proportion,
which at present is 12%. In Finland, progress has been faster,
21% of professors being female.

The main reason for the successful development in Finland has
probably been the very active education of biomedical scientists,
which was initiated in the mid-1990s (Academy of Finland
2003). The Finnish Ministry of Education decided to create a
graduate school system, and the first four-year schools started
in 1995. The amount of governmental spending on PhD education was considerably increased, while additional funding
for the students also came from other sources, such as private
foundations and industry.

During this programme, the proportion of PhD degrees
awarded to women rose to 46% (including technology). In

medicine, half of the students achieving PhD degrees in 2003 have been female. In the public sector and in the universities, women accounted for over 40% of the research staff in 2001, but in business and industry the figure was only 20% (Academy of Finland, 2003). Hence, governmental funding of education and research appears to be a good way to increase the proportion of female scientists.

Governmental funding does not alone guarantee that it is fairly distributed. Wennerås and Wold (2000) analysed the proportions of female and male recipients of grants from the Swedish Research Council. They found significant under-representation of the female applicants among those who were awarded grants. They analysed the scientific work on which the applications were based, and found that a female applicant had to have a much higher total impact factor score in her preceding publications than her male colleagues in order to obtain a grant. The difference was comparable to the impact of one article in the journal *Nature*. This appears to happen in other countries too (Mavis and Katz 2003).

Since the study by Wennerås and Wold, the Swedish Research Council has actively paid attention to equity and, as a consequence, the proportion of new grants to women increased from 30% in 2000 to 45% in 2003 (www.vr.se). The proportion of female members in the evaluating committees increased from 10% in 1996 to 40% in 2002. This certainly has had an impact on female scientists' work in Sweden.

There are others factors to consider when thinking of optimal working conditions for female scientists. High quality day-care facilities should be available for families, hence allowing young parents to be optimally educated in research. This would also help women to have their children at the physiologically optimal age. Optimal day-care would help young parents to be employed without the threat of discrimination. When awarding research grants, the time spent on maternity or paternity leave should be excluded from research time when scoring the productivity of individuals.

Discrimination against pregnant women when appointing people to posts is already illegal in most European countries, but there are still hidden attitudes that influence the possibility of a pregnant woman being appointed. Local, national and international women's networks are still of great importance for women's positions in science.

The ethical principles regarding study subjects are similar for women and men. The main concern regarding women as study subjects is the possibility of pregnancy and the effects of the trials on foetal development. The risks of pharmaceuticals and chemicals have been recognised, and there are harmonised international regulations concerning how these substances have to be tested before any trials. The concern over foetal effects is why pregnant women particularly are a very specific group of study subjects.

There are clear restrictions, which, however, have impaired the gathering of knowledge of certain aspects of women's health. It has been easier to have male study subjects. Female scientists are more likely to promote the issues particularly important for women. Women are under-represented in leading positions in research. All possible ways to promote the careers of female scientists are extremely important.

Bibliography

Academy of Finland 2003. PhDs in Finland : *Employment, placement and demand.* Publications of the Academy of Finland, 5/03. www.aka.fi

EGE (the European Group on Ethics in Science and New Technologies to the European Commission) 2000. *Ethical aspects of human stem cell research and use,* European Commission No. 15, 14 November 2000, rapporteurs Anne McLaren and Göran Hermerén.

EU 2002. *More Research for Europe : Towards 3% of GDP,* Communication from the Commission, Brussels 11.9.02, COM.

EU 2003. *Third European Report on Science and Technology Indicators 2003 : Towards a Knowledge-based Economy,* European Commission, Directorate-General for Research – K – Knowledge-based economy and society.

Mavis, B. and Katz, M. 2003, "Evaluation of a programme supporting scholarly productivity for new investigator." *Academic Medicine,* 78 : 757.

Thomson, J.A., Itskoviz.t-Eldor, J., Shapiro, S.S. et al. 1998. "Embryonic stem cells derived from human blastocysts", *Science* 282 : 1142.

Wennerås, C. and Wold, A. 2000. "A chair of one's own", *Nature,* 408 : 647.

Biomedical research in Europe

Germany: current legislation

by Jochen Taupitz

The following chapter describes the legal situation in Germany as of August 2003. Since Germany has not yet acceded to the Council of Europe's European Convention on Human Rights and Biomedicine (the Bioethics Convention), its provisions are not considered in the following chapter, although significant differences are noted.

Applicable rules of law

Germany has no comprehensive law on research or on the protection of patients or persons used in scientific research. Rather, German law attempts to cover certain relatively easily definable risks in biomedical research with special sets of rules that supplement the general provisions of civil, criminal and public law.

In Germany, legislative power is divided between the Federation and the *Länder*, but most of the regulations on the dangers of biomedical research are federal. The *Land* regulations, on the other hand, relate principally to specific professions, due to the fact that the *Länder* have exclusive competence to legislate on the exercise of professions, especially the healing professions.

Special federal regulations on research on human beings

The following are the main areas regulated at federal level:

- the clinical examination of drugs (Sections 40ff. of the Drugs Act, in the version of 11 December 1996);[1]

- the clinical examination of medical products (Sections 20ff. of the Medical Products Act, in the version of 7 August 2002);[2]

- the hyper-immunisation of test subjects (Section 8 of the Transfusion Act, in the version of 1 July 1998);[3]

- research with radioactive substances and ionising radiation (Sections 33ff. of the Radiation Protection Order, in the version of 20 July 2001);[4]

1.
BGBl (Federal Gazette) I 1998, p. 3586, last amended on 21. 8. 2002, *BGBl.* I 2002, p. 3348. Directive 2001/20/EC of the European Parliament and of the Council of 4 April 2001 on the approximation of the laws, regulations and administrative provisions of the Member States relating to the implementation of good clinical practice in the conduct of clinical trials on medicinal products for human use (OJ No. L 121 p. 34 of 1.5.2001) has yet to be incorporated into German law. At present, there is only a ministerial draft.

2.
BGBl. I 2002, p. 3146.

3.
BGBl. I 1998, p. 1752.

4.
BGBl. I 2001, p. 1714, last amended on 18.6.2002, *BGBl.* I 2002, p. 1869.

- the import and use of human embryo stem cell lines in the Stem Cell Act of 28 June 2002[1] – although the Protection of Embryos Act of 13 December 1990,[2] which protects the embryo all but absolutely until its nidation in the uterus, also imposes considerable restrictions on research on germ cells, germ line cells, embryos and embryo cells.

Land regulations on exercise of the medical profession

The *Länder* also have legal regulations, particularly on the exercise of the medical profession (including research), which apply in cases where the above more specific regulations do not. In some cases, they have separate laws on doctors, dentists, pharmacists, etc. In other cases, they have laws on professional associations, covering several health professions. All of these *Land* laws give professional associations under public law (membership of which is compulsory for persons exercising the professions concerned) the competence to issue professional codes in the form of public-law regulations.

Some of these regulations also contain rules on biomedical research. They stipulate, for example, that an ethics committee must be consulted before an experiment begins (cf. Article 15 of the Model Professional Code, produced by the Federal Medical Association as a basis for the codes of the *Land* medical associations).

The World Medical Association's Helsinki Declaration has also had a powerful influence on the profession, although it has no legal force, and compliance is voluntary. Moreover, provisions concerning research on persons incapable of giving consent in the new version (2000) have been severely criticised in Germany.

Finally, mention should be made of the law applied to universities, which imposes on their members public-law obligations of varying scope in the matter of research on human beings.

General regulations

1.
BGBl. I 2002, p. 2277.

2.
BGBl. I 1990, p. 2746, last amended on 23.10.2001, *BGBl.* I 2001, p. 2702.

Outside those areas of practice where special laws apply, it is largely unclear how research should be assessed in German law. It is often argued that these special laws merely express general legal principles, and can thus be applied *mutatis mutandis* in areas they do not specifically govern – but this does not explain the reason why the legislature has simply enacted (non-uniform) rules on a few areas, instead of formulating general legal rules on research on human beings, or at least rules applying to all the specialised areas concerned. A better approach would be to try to lay down legal principles for medical research on human beings (which might also have been embodied – at least partially – in the special laws).

Nonetheless, research conducted outside the purview of the special laws does not take place in a legal vacuum. The legal text that outranks all others is the Basic Law (*Grundgesetz*, the German Constitution) – especially where the fundamental rights of patients and research subjects, and of researchers, are concerned. Most legal experts take the view that the fundamental rights protected by the Basic Law are not directly applicable to research conducted by non-state authorities, since the Basic Law is concerned with the individual's rights vis-à-vis the state (and not vis-à-vis other individuals).

However, those fundamental rights also shape and influence the rules of civil law, and so have an indirect effect on legal relations between private individuals. In addition to protection of human dignity (Article 1, paragraph 1 of the Basic Law), the basic rights most relevant to patients and research subjects are the right to free development of personality, including general personality rights (Article 2, paragraph 1), the right to life and physical integrity (Article 1, paragraph 2.2) and the right to self-determination derived from the others (Article 2, paragraph 2.1 in conjunction with Article 1, paragraph 1). The rights most relevant to researchers are freedom of research (Article 5, paragraph 3), which can be limited only by the basic rights of others or by other constitutionally protected interests, freedom to exercise a profession (Article 12, paragraph 1) and general freedom of action (Article 2, paragraph 1).

As for ordinary law, the general law of contract and tort, and general criminal and administrative law, also apply to biomedical research. Specifically, this means that offences involving bodily harm, as defined in Article 823(1) of the Civil Code and Articles 223ff. of the Criminal Code, are relevant – as is the protection of general personality rights (included under "other rights" in Article 823(1) of the Civil Code) in civil law. Also relevant is liability for violations of protective laws, under Article 823(2) of the Civil Code.

Other applicable provisions include the rules on doctors' obligation to maintain secrecy and the data-protection regulations of the Federation and the *Länder*, some of which have regulations which make it easier to use data for research than for other purposes.

Research institution guidelines

The guidelines of the major German research institutions (for example, the German Research Foundation) also do much to ensure that the research they fund respects certain quality standards, and that patients and research subjects are protected. For example, the German Research Foundation will not fund research on human beings without the consent of of an ethics committee.

Requirements for biomedical research on human beings

Separation from clinical practice

Biomedical research and clinical (medical) practice must be separated. Clinical practice, in which the doctor–patient relationship is regulated partly by medical association statutes, but mainly by judge-made law, comprises standard and individual treatment. Both are exclusively focused on the individual patient's well-being, but one is newer than the other. On the other hand, biomedical research – as well as being innovative – involves systematic, research-orientated planning and systematic evaluation of results.

In other words, it also sets out to extend medical knowledge beyond the individual case, that is, to serve the common good. Since patients (people who are actually suffering from the disease on which research is being done) and research subjects (people who are not suffering from that disease) are partly subjected to measures during the research that do not benefit them directly, and are often exposed to certain risks in the process, it is recognised that they need special protection.

Basic types of biomedical research

Biomedical research includes both therapeutic research and scientific experiments. As well as the advancement of scientific knowledge, therapeutic research is designed and intended to improve the health of the participating patient. It is thus both possible and necessary to make an assessment, including not only the gains in scientific knowledge, but also – and above all – the ratio of benefits (expected) to risks (feared) for the patient. Scientific experiments, on the other hand, seek merely to add to scientific knowledge, and this means comparing two things – general gains and individual risks – which are not really comparable.

The (admittedly fluid) distinction between therapeutic research and scientific experiment is important in several areas, of which the chief are :

• *Risk/benefit analysis:* In the case of a measure taken in the patient's direct interest, a very high level of risk may be acceptable (for example, life-threatening cardiac surgery, when this is the patient's only hope). On the other hand, a measure that mainly or solely benefits the community should be attended by minor risks only.

• *Presumed consent:* Consent may indisputably be presumed in the case of measures which are objectively in the patient's interest, but not in the case of experiments that simply contribute to scientific knowledge. This is why some authors take the view that scientific experiments cannot be based on presumed consent. However, this is probably going too far. The most that can be said is : the less clearly a measure is in the patient's objective interests, the stronger must be the reasons for presuming consent.

• *Research with people incapable of giving consent:* There is considerable disagreement as to whether scientific experiments with people incapable of giving consent should be totally banned, or whether experiments of merely possible benefit to the person concerned, and experiments that do not benefit him/her, but will – it is hoped – benefit others in the same group, may sometimes be permitted. The Drugs Act (*Arzeimittelgesetz*) and the Medical Products Act (*Medizinproduktegesetz*) permit research (only) with diagnostic and prophylactic drugs and medical products, including research on healthy minors (that is, research with possible benefits only for the person concerned), but not on adults incapable of giving consent.

Freedom of research: protection of patients and test subjects

In principle, Article 5, paragraph 3 of the Basic Law guarantees freedom of research, and thus of biomedical research as well. In other words, research requires no special legal permission. However, it is clear from the rules outlined above that there are – in the interests of patients and research subjects – numerous restrictions on research. However, because there are, as we have said, so many regulations, there is no fixed catalogue of protection criteria, applying equally to all research projects. Moreover, the content of the various regulations sometimes differs considerably. The following is thus a mere overview.

Protection criteria can be roughly divided into three groups: objective criteria, criteria designed to protect the individual's right of self-determination, and procedural safeguards. Objective criteria include risk/benefit analysis and requirements concerning the qualifications of the researcher. The main criterion concerned with the individual's right of self-determination is the requirement that his/her consent (or that of his/her legal representative) must be obtained. Finally, procedural safeguards include the involvement of independent outsiders (especially ethics committees) or the obligation to submit the details to a public authority.

Overview of protection criteria

Informed consent

It is generally agreed that research involving human beings is permissible only with the consent of the person concerned. To be valid, consent must be informed – that is, the person concerned must be given a clear picture of the nature, significance and consequences of the measure, including any risks. Consent may be revoked at any time with future effect. This is expressly stated in various special laws, but the Transfusion Act probably takes this point for granted and so does not mention it explicitly. All the special regulations we have mentioned call for written consent – advisable in any case for purposes of proof.

For consent to be valid, the person concerned must also be capable of giving it. A person is capable of giving consent if he/she can at least roughly understand the nature, significance and consequences of the measure, weigh up the pros and cons and reach an informed decision on this basis. Individual circumstances must be considered here. The predominant view is that the age qualifications which determine the capacity of minors to enter into legal transactions (limited capacity from the age of 7, full capacity from the age of 18), are not applicable. In practice, this is a source of considerable uncertainty.

If the person concerned is unable to give consent, the decision is taken (once they have been duly informed) by their legal representative (normally the parents for a minor, the statutory guardian for an adult). Under ordinary family law, the legal representative is required to consider the protected person's "welfare". There is considerable disagreement regarding the scope of this term, although the prevailing view is that at any rate, it must not be taken to mean "physical well-being" only.

A person may also be represented by an authorised agent, meaning someone to whom they have granted the right to handle their affairs. However, this has rarely happened in the field of medical research.

In the special laws, the situation of adults unable to give consent, and of minors able or unable to give consent, is regulated in various ways:

- All the laws allow therapeutic research but, in the case of minors legally unable to give consent, in addition to the consent of their legal representative, their own must nonetheless be obtained.

- Both the Drugs Act and the Medical Products Act prohibit scientific experiments with adults unable to give consent. Their legal representative's consent makes no difference.

- As already stated, the Drugs Act and the Medical Products Act allow diagnostic and prophylactic drugs, and medical products to be tested on minors, if these drugs/medical products are used to detect or prevent children's diseases, and their use is justified, according to medical knowledge, in order either to detect diseases in the minor concerned or to protect him/her against them. In the case of minors legally capable of giving consent, both the consent of their legal representative and their own is again required.

Informed consent is not required if, according to medical knowledge, the research in question is justified to save the patient's life, restore their health or alleviate their suffering; provided that informing them would jeopardise the treatment's success, and that no objections are discernible. Some of the special laws cover this point expressly, but the Radiation Protection Order and the Transfusion Act do not mention it.

Risk/benefit analysis

Consent does not always legitimise research. Rather, the risks involved for the person concerned must be medically justifiable when set against the likely benefits. Unlike the Council of Europe's Convention on Human Rights and Biomedicine, German law does not generally require more stringent risk/benefit analysis in the case of research carried out for the benefit of others.

Official procedure

All the special laws provide for an official procedure for conducting research, but only the Radiation Protection Order and the Stem Cells Act insist on prior authorisation. Under the other laws, it is sufficient to submit the relevant documents to the authorities or to give them notice of the project – thus allowing them to prohibit it, if necessary.

Ethics committees

Under all the special laws and the law on the medical profession, research projects must first be examined by an independent and interdisciplinary ethics committee. However, there are differences as to:

- whether the examination must be carried out by an ethics committee under public law (Drugs Act, Transfusion Act, the law on the medical profession), or whether consulting a (registered) ethics committee is enough (Medical Products Act, Radiation Protection Order);

- whether consultation in itself is sufficient, for example even an advisory verdict rejecting the proposal, (Radiation Protection Order, the law on the medical profession), or whether a positive approval is required (Drugs Act, Medical Products Act, Transfusion Act). However, under the Drugs Act and the Medical

Products Act, research can even begin without a positive approval from the ethics committee, if the supreme federal authority concerned does not object to the project within sixty days of receiving the documents;

- whether one verdict is sufficient for multicentre projects (Medical Products Act) or whether local ethics committees in all the areas where the project takes place must be consulted (Drugs Act, Radiation Protection Order, Transfusion Act, the law on the medical profession).

The Stem Cells Act established a special Central Ethics Committee for Stem Cell Research, which is the only ethics committee operating at federal level. In other fields, there are no central (that is, national) ethics committees.

At present, there are no uniform procedural rules or guidelines on the composition of ethics committees, and standardisation is therefore badly needed.

Assessment plan

Under all the special laws, the research before the authority or ethics committee concerned must be based on a scientific assessment plan.

Obligation to notify unexpected occurrences

Under the Drugs Act and the Transfusion Act, the ethics committee must be informed of unexpected events.

Qualifications of clinical trial directors

The Drugs Act, the Medical Products Act, the Radiation Protection Order and the Transfusion Act all lay down special requirements for the qualifications of clinical trial directors:

- Drugs Act: a doctor with at least two years' experience in clinical testing of drugs;

- Medical Products Act: a specialised doctor with appropriate qualifications. In the case of medical products for use in dental treatment, there must also be a dentist or some other suitably qualified and authorised person with at least two years' experience in clinical testing of medical products;

- Radiation Protection Order: a doctor who has at least two years' experience of using radioactive substances or ionising radiation on human subjects, has the necessary radiation protection expertise and can be reached at any time during the research. An expert on medical physics must also be involved in the planning and implementation phase;

- Transfusion Act: a registered physician familiar with the latest developments in this field.

Necessity or quality of research

The Radiation Protection Order, the Transfusion Act and the Stem Cells Act lay down certain requirements concerning the necessity or quality of research.

Under the Radiation Protection Order, there must be a compelling need for the project because previous research findings and medical knowledge are insufficient. It must also be impossible to replace the use of radioactive substances or ionising radiation with a type of examination or treatment which does not expose the subject to radiation. Moreover, the radioactive substances or ionising radiation used in the research must match their purpose. It must be impossible to replace them with other substances or applications that involve less exposure for the research subject. Finally, it must not be possible, in the present state of medical knowledge, to reduce exposure to radiation or the activity of the substances used any further without jeopardising the purpose of the project.

Under the Transfusion Act, the donor immunisation needed to obtain plasma for the production of special immunoglobulins may be carried out only when, and for as long as, this is necessary to ensure adequate stocks of these products, and in accordance with the latest medical knowledge and technology.

Under the Stem Cells Act, research using embryonic stem cells may be carried out only if it has been scientifically shown that:

- its findings will make an important contribution to scientific knowledge and be usable either in basic research or in developing new diagnostic, preventive or therapeutic methods for use on humans;

- in the present state of science and technology, the questions to be studied have already been investigated, as far as this is possible, with the help of *in vitro* models using animal cells, or animal experiments;

- the scientific knowledge hoped for from the project can foreseeably be obtained only by using embryonic stem cells.

Unlike the Bioethics Convention, none of the special laws stipulates that there must be no alternative to research on human beings, or expressly states that the interests of the individual must always take precedence over the common good.

Prior safety tests

All the special laws require the carrying-out of certain safety tests (for instance, pharmacological and toxicological tests) before research is conducted.

Provision for injury

Under the Drugs Act, the Medical Products Act and the Radiation Protection Order, patients and test subjects must be compensated for any injuries, regardless of whether fault is involved. Some of those involved in the politico-legal debate insist on compensation regardless of fault, for research out of the application area of the special laws mentioned above.

Specific groups

Under the Drugs Act, the Medical Products Act and the Radiation Protection Order, research may not be carried out on persons detained in an institution by order of a court or public authority.

The Medical Products Act and the Radiation Protection Order contain special protective rules on research during pregnancy and lactation.

When research using persons unable to give consent is permitted by the special laws (see page 113) or is generally considered to be acceptable, there is still a requirement that research using persons able to give consent must be thought unlikely to yield adequate results. Research with the latter must always be preferred to research with the former.

There is no general ban on improper attempts to influence research subjects.

The following are the main issues currently under discussion in Germany:

• Expansion of research with persons unable to give their consent.

• Expansion of research with embryos and embryonic stem cells.

• Limits on human genetic research, including research with human body parts.

• Greater financial involvement of public health insurance bodies in research.

Bibliography

Deutsch, E./Spickhoff, A. 2003. *Medizinrecht,* 5th edn; Berlin, Springer.

Taupitz, J. 2002. *Biomedizinische Forschung zwischen Freiheit und Verantwortung;* Berlin, Springer.

Central and eastern Europe: research-related problems for transition countries

by Eugenijus Gefenas

It is a challenge to talk about central and eastern Europe (CEE) as a whole without lapsing into oversimplified generalisations or neglecting the peculiarities of individual states. The main reason for making this reservation is the diversity of socio-economic conditions and cultural backgrounds in the various transition countries in the region. For example, some of the CEE countries are still taking their very first steps towards democracy and, in addition, have to cope with an extremely difficult economic situation. That is why my observations will be mainly based on the Baltic countries' experience, which will, I hope, also cover common problems arising in the course of ethically sensitive biomedical research in other European transition societies.

Establishing research ethics committees in central and eastern Europe

Let me first make a few observations about the motives for setting up research ethics committees (RECs) and meeting the requirements of research ethics in the CEE countries, and the incentives to do so. It could be argued that the initial incentive – to meet the formal requirements of an ethical review of biomedical research comparable to that in western countries – was introduced in many central and eastern European countries by foreign pharmaceutical companies (Simek et al. 2000).

It was most probably because of the willingness of CEE researchers to take part in multi-centre clinical trials that the intensive process of establishing RECs started in the most scientifically active healthcare research institutions in the late 1980s and early 1990s. It also explains why research ethics committees have been predominantly based in teaching hospitals and therefore came to resemble the institutional review boards operating in the USA (Trontelj 2000) rather than the regional RECs of the Scandinavian countries. Poland is a good example of this tendency, because in this country each medical school has its committee for research involving human subjects (Górski and Zalewski 2000).

It is worth pointing out that the procedural requirement to apply for and obtain approval from the local REC before starting any biomedical research project has been recognised in almost all the countries of central and eastern Europe. More importantly, however, in order not only to follow the procedural rules of ethical review but also to implement the substantive principles of research ethics in practice, it is necessary to be sensitive to those cultural and socio-economic features of a particular society that have a negative impact on the freedom of choice of the subjects who are to participate in biomedical research.

That is why I shall concentrate on two main features that tend to increase the vulnerability of the research population in central and eastern Europe. These features have been linked with the scarcity of resources in the healthcare sector as well as with a somewhat insufficient culture of respect for personal autonomy as regards health care.

Financial incentives and the vulnerability of research participants

A traditional list of vulnerable research subjects includes, among others, children and adults with behavioural disorders, as well as those whose voluntary choice as regards participation may be restricted by their dependency on particular institutions or persons. The concept of vulnerability might also be extended to cover participants from less developed and less affluent societies, who may be vulnerable because of their lack of sophistication when it comes to modern scientific medicine and because their financial status may be exploited by researchers (Brody 1998).

Research as a means of obtaining treatment

Let us therefore first analyse the economic context of the ethical review of biomedical research in central and eastern Europe. We need to be aware that, in the majority of the former Eastern Bloc countries, annual per capita healthcare expenditure is on average five to ten times lower than in many welfare-state societies. For example, it amounts to US$250 in Lithuania, which might be regarded as a medium-income country in the wide spectrum of CEE states, as compared with US$ 1 700 in Sweden (Brody and Lie 1993). Sophisticated biomedical technology and expensive drugs available in the more affluent country would not therefore normally be accessible in many transition societies, not to mention the fact that the poorest CEE states face a shortage of quite basic biomedical material, such as disposable syringes and antibiotics.

That is why the argument that a clinical trial provides an opportunity for the research participants (or some of them at least) to receive treatment not otherwise available will receive strong support in these circumstances. This state of affairs might also lead to a situation whereby a trial that had not been approved in a Western country would be regarded as acceptable in the transition society.

To illustrate the problem, it is worth recalling the well-known example of the AZT* placebo-controlled trial on pregnant woman conducted in Africa in the 1990s. The trial was designed to establish the effectiveness of a smaller dosage of AZT in preventing HIV in newborns and involved two groups of pregnant women, the first receiving the active medication, the second one being a control group given a placebo. At that time, such a trial would not have been approved by the ethics committees in any developed country in the world because the effectiveness of AZT had already been established.

This trial revealed the double standards of research ethics in the developed and the developing world. However, its proponents argued that it was justified in Africa, where the drug was not available at all and nobody was therefore worse off as a result of the trial, while at the same time the research was beneficial to the group of participants receiving the medication.

It is important to note, in this connection, that a few international instruments have recently addressed the issue of protection of research subjects in societies with limited resources. For example, CIOMS Guideline 10 expressly states that, before research is transferred from affluent countries to populations with limited resources, the investigator and sponsor should be responsive to the health needs and priorities of the communities or populations concerned and, furthermore, should try to make the results of the research "reasonably available" to them (CIOMS Guidelines 2002).

AZT:
an antiviral drug (trade name Retrovir) used in the treatment of Aids; adverse side-effects include liver damage and suppression of the bone marrow.

Research as a source of income

Another important aspect of the shortage of financial resources conducive to the vulnerability of the research population is related to low salaries in the healthcare sector. Researchers in the region have a relatively stronger incentive to conduct clinical trials proposed by pharmaceutical companies because the benefits offered in the transition societies are relatively much higher than in Western countries.

For example, the payments received by the researchers in remuneration for conducting a clinical trial could very well exceed their regular salaries, especially if we add such hidden types of remuneration as the reimbursement of expenses for overseas conferences and the like. This argument carries the most weight in the CEE countries with the largest difference between the income of healthcare practitioners and the remuneration offered by the pharmaceutical industry. Consider, for example, certain former Soviet republics where the doctor's official monthly salary hardly exceeds US$50, while the fee for the recruitment of the research subject amounts to a few hundred US dollars.

Combined with the tradition of medical paternalism, the financial incentive to enrol research participants makes it very likely that some of the basic principles of research ethics may not be followed in the circumstances. At European level this issue has recently been addressed by the Council of Europe's Protocol on Biomedical Research to the Convention on Human Rights and Biomedicine. Article 12 of the protocol stresses the need to protect vulnerable and dependent persons against "undue influence", which might be exerted on persons to participate in research (Additional Protocol on Biomedical Research to the Convention on Human Rights and Biomedicine, 2004).

The paternalistic tradition and the vulnerability of research subjects

The paternalistic doctor–patient relationship prevalent in post-communist societies is another important issue for research ethics. It leads to a situation whereby patients simply do not ask the questions that are essential if they are to give their free and informed consent and, in addition, are not able to resist the proposals made by the doctor/researcher. In other words, a biased researcher might easily exploit a paternalistic attitude and consequently mislead a patient, creating false expectations as to the main purposes of the clinical trial in which he or she is participating.

To explore all these important issues and evaluate the practical implications of informed consent, the Lithuanian Bioethics Committee conducted an anonymous postal survey of participants in clinical trials carried out in the country during the year 2001. Of 1 106 questionnaires sent to the research participants, 438 completed questionnaires were returned to the committee, revealing a number of problems related to the practical implementation of the principle of informed consent (Lukauskaite 2003).

Let us now concentrate on one particular aspect of the study, which I would consider a test case as regards informed consent, and which entailed checking whether research subjects participating in the placebo-controlled double-blind* (PCDB) research project had understood the design and essential features of such a trial.

Placebo-controlled double-blind clinical trial: a "test case" for informed consent

About 100 anonymous questionnaires were received from the participants in PCDB clinical trials. The evaluation of the quality of informed consent to this type of trial was based on checking the participant's understanding of three essential features:

- the main goal of the trial;

- the use of a placebo;

- the double-blind design.

The PCDB trials could hardly be regarded as being conducted with the intention of benefiting individual patients. However, the question "What do you think is the main goal of the doctor conducting this clinical trial?" revealed that almost a half of

Double-blind study:
a clinical trial design in which neither the participating individuals nor the study staff knows which participants are receiving the experimental drug and which are receiving a placebo (or another therapy). Double-blind trials are thought to produce objective results, since the expectations of the doctor and the participant about the experimental drug do not affect the outcome; also called double-masked study.

the participants had a "therapeutic misconception" about the research as they thought that the aim of the trial was to improve their – or their child's – health.

The fact that some of the participants in the PCDB trial receive a new drug while others receive a substance without any pharmaceutical activity (a placebo) was understood by only one-third of the research participants. Finally, the question "Do you think your doctor has been aware of what exactly is being administered to you?" was answered positively by almost 40% of participants, who evidently misunderstood the double-blind character of the trial.

The results of the survey give cause for concern about respect for the principle of informed consent as well as for the safety of the patients. Such an impression, however, needs to be qualified. Firstly, the survey covered only international, multi-centre clinical trials conducted simultaneously in many different European countries. The placebo group was a justified feature of these trials as it was only used where no effective treatment was available and/or where giving a placebo resulted in minimal risk and burden. Secondly, the misunderstanding of the way in which the PCDB trial was designed could also be related to the fact that some research subjects would find it difficult to follow relevant information even if this information were presented in a completely adequate way by the researcher.

The national genome project

By turning to molecular biology and genetics, modern medicine has held out greater promise of "personalised" medicine and pharmacogenetics. In Estonia and some other countries, this process has even sparked off national population-based genome projects. The Estonian project is probably one of the most developed and well-known in the Baltic region and Europe generally. Because information about the project (including the information given to potential participants) has been made available on the Internet, it might serve as a good "test case" for research ethics, with particular reference to its features in the transition societies.

The Estonian project has been praised as having the potential to become Estonia's Nokia – the most promising facilitator of scientific and even economic development in the country. At the same time, because of the lack of informed public debate about it and the potential for future abuse, the project has been called "an ethical time bomb" not only by some local commentators but also by international commentators (Gross 2000). Even though this might be regarded as going too far, we have to be aware that genetic research is a highly controversial field.

It seems that the ethical controversies revealed by the Lithuanian survey of informed consent to pharmaceutical trials are also applicable to the genome project. Let us first consider its financial context, in particular the incentives for the general practitioners who are supposed to serve as recruiters of the participants in the genome research. One of the leaders of the Estonian Genome Project was quite eloquent in commenting on the economic advantages of conducting the project in Estonia. He estimated that Estonian general practitioners would be willing to work for US$10-15 per patient, whereas in a country such as the USA that amount of money would not even allow one "to walk through the door" of the doctor's surgery (Mapping Estonia, www.internationalspecialreports.com/europe/01/estonia/education).

This argument, however, can be viewed from a different angle. Even though the same US$15 might indeed be regarded as a sufficient incentive to work for the project, this sum could also be seen as having the potential to elicit abuse. Recruiting two or three patients per day would almost double the income of the Estonian doctor. This might also strengthen our misgivings as to how free to choose and how well informed those who sign the genome project participation documents in their GP's clinic actually are, given the "therapeutic promises" made in the gene-donor consent form.

The problem is that the "non-therapeutic" features of the project do not seem to be clearly explained to the research participants. Describing possible benefits, the information accompanying the gene-donor consent form merely says that "The Gene Bank provides a gene donor with an opportunity to assess his or her health risks and diagnose illnesses more precisely, prevent falling ill and receive more effective treatment in the future."

(Gene donor consent form, 2001). How distant this "future" might be, and how uncertain the process of finding effective treatment for genetic disorders still is, remain unexplained in the written information documents given to the patient. Perhaps it is no coincidence that the gene-donor consent form lists the right not to know one's genetic data before the right to know one's own genetic data.

Besides the criticisms of the features of the genome project mentioned above, there are some more complicated problems to be highlighted in this context. Is it at all possible to run an Estonian genome project on the same model of informed consent as is applied in case of an ordinary PCDB pharmaceutical trial? The problem is that this project has a very broad and ambitious aim, namely "genetic and medical research to be carried out in order to find genes that influence the development of illnesses" (Gene donor consent form, 2001).

Such a broad goal could hardly be consistent with the fundamental criterion of informed consent, namely, that sufficiently specific and explicit information is provided about its future relevance to the individual concerned. However, it is important to note that even if the future research use of human genetic material and personal data cannot be precisely anticipated, "unconditional blanket consent" should be avoided as it cannot be seen as a valid form of informed consent (Paragraph 64 of the *Draft explanatory report to the draft instrument on the use of archived human biological material in biomedical research,* 2002).

Future measures

What are the ways of coping with the situation when people are enrolled for different types of biomedical research without actually being aware of what kind of procedures are being offered to them and what potential benefits, if any, they might realistically hope for?

One suggestion might be more active monitoring of biomedical research activities. The anonymous survey of respect for the principle of informed consent to pharmaceutical trials could serve as an example of a means of establishing feedback from research participants to researchers. For example, the results of the anonymous Lithuanian survey of informed consent were presented to the researchers at a special conference, and triggered significant interest among the biomedical community.

This example of an anonymous survey might also be interesting from an international perspective. Even though the difficulties related to the practice of informed consent are greater in the countries with a relatively short history of research ethics, it might also be useful to compare compliance with the fundamental principles of research ethics in different sociocultural contexts, for example, not only in CEE countries but also in southern, western and northern Europe. The globalisation of biomedical research and the increasing number of multi-centre clinical trials make this idea quite feasible. We should, however, be very careful to prevent such monitoring being transformed into policing.

The controversy surrounding the genome studies might prompt us to direct our efforts to somewhat broader research ethics issues. These projects reveal more systematic difficulties in conforming to the standards of informed consent applied in traditional clinical trials and might therefore also prompt us to think about alternative research ethics paradigms. For example, some bioethics experts would argue that it might not always be feasible to obtain informed consent, even from individuals who are competent to give it and understand the issues, because it is not possible to foresee the full range of uses to which genetic information might be put. According to the proponents of this point of view, the strict requirement of informed consent evolved as a reaction to Nazi experiments, whereas the context has changed in contemporary research. An alternative principle of solidarity might be proposed in the case of research with so-called "minimal risk" (Chadwick and Berg 2001).

Both researchers and those responsible for ethical review therefore find themselves in a rather difficult situation, where the established rules of biomedical research cannot be directly applied and there is an urgent need to look for alternatives or exceptions in the case of genetic research or research on archived human biological material and personal data. The situation is more confusing in the CEE countries, where not even the standard rules of research ethics have yet been fully established.

In spite of the critical spirit of my observations, I have to stress that, since the collapse of the Eastern bloc in the late 1980s, many post-Soviet European countries have made remarkable progress in the field of biomedical research ethics. The majority of countries have already established research ethics bodies and are on their way to introducing an efficient system for the ethical review of biomedical research.

This paper should not, therefore, be seen as a pessimistic evaluation of the processes taking place in the European countries in a state of transition. It should, rather, be regarded as a sign of increasing transparency and openness to positive changes. The very fact that information about ongoing research activities and surveys of these activities are becoming accessible to the public is the main guarantee of successful future developments.

Bibliography

Brody, Baruch 1998. *The Ethics of Biomedical Research: An International Perspective*, Oxford, Oxford University Press, p. 50.

Brody, Baruch and Lie, Reidar 1993. "Methodological and conceptual issues in health care system comparisons: Canada, Norway, and the United States", *Journal of Medicine and Philosophy*, vol. 18, pp. 437-63.

Chadwick, Ruth and Berg, Kare April 2001. "Solidarity and equity: new ethical frameworks for genetic databases [on line]", *Macmillan Magazines*, vol. 2, available on the Internet: http://www.nature.com/.

CIOMS. *International Ethical Guidelines for Biomedical Research Involving Human Subjects*, 2002, CIOMS, Geneva, pp. 51-3.

Draft explanatory report to the draft instrument on the use of archived human biological materials in biomedical research, Steering Committee on Bioethics (CDBI), CDBI/INF (2002) 6, Strasbourg, 14 October 2002), available on the Internet, http://www.coe.int/bioethics.

Gene donor consent form [online], Regulation No. 125, 17 December 2001. Estonian Minister of Social Affairs, available on the Internet: http://www.geenivaramu.ee

Gross, Michael 2000. "Estonia sells its gene pool" [online], *The Guardian*, 9 November 2000, available on the Internet: www.guardian.co.uk/science/story (17.08.2002).

Górski, Andrzej, and Zalewski, Zbigniew, 2000. "Recent developments in bioethics in Polish science and medicine" in *Ethics committees in central and eastern Europe*, ed. J. Glasa, IMED Fdn., Bratislava, pp. 209-15.

Lukauskaite, Kristina, "Informed consent in Lithuanian clinical trials: problems and possible solutions", Proceedings of the seminar "Informed consent: from theory to practice", 6 May 2003, Vilnius, published by the Lithuanian Bioethics Committee (in Lithuanian).

Mapping Estonia : Estonia takes an ambitious project to outline the country's genome [online]. Education. Available on the Internet : http ://www.internationalspecialreports.com/europe/01/estonia /education/.

Protocol on Biomedical Research to the Convention on Human Rights and Biomedicine [online], 2003, Council of Europe, available on the Internet : http ://www.coe.int/bioethics.

Simek, Jiri et al. 2000. "Ethics committees in the Czech Republic" in *Ethics committees in central and eastern Europe*, ed. J. Glasa, IMED Fdn., Bratislava, pp. 125-31.

Trontelj, J. 2000. "Ethics committees in Slovenia" in *Ethics committees in central and eastern Europe*, ed. J. Glasa, IMED Fdn., Bratislava, 2000, pp. 239-50.

Italy: some shortcomings of biomedical research

by Stéphane Bauzon

The legal situation of biomedical research in Italy is ambiguous. On the one hand, Article 32 of the constitution (concerning the individual's fundamental right to health), which stipulates that no one may be obliged to take part in biomedical research, seems to be fully complied with: Italy has legal rules that strictly protect biomedical research subjects. On the other hand, Article 9 of the Constitution, which states that "[t]he Republic shall promote the development ... of scientific research" has by no means been translated into practice. The high degree of protection afforded to all biomedical research subjects therefore contrasts with the small amount of medical research.

Italy obviously endorses the World Medical Association's 1964 Declaration of Helsinki (which concerns the ethical principles applicable to medical research on human subjects). The Declaration has been amended at the World Medical Association's various general assemblies, one of which (the 35th) took place in Venice, Italy in 1983. Italy recognises the universal importance of the Declaration of Helsinki and its fundamental principles, for example:

- The principle of respect for the individual, which covers the capacity and right of all individuals to make their own choices and decisions. It concerns respect for the independence and self-determination of all human beings, whose dignity and freedom are acknowledged. One important aspect of this principle is the special protection that needs to be afforded to vulnerable people.

- The principle that benefits must be derived from the research, which requires that researchers assume responsibility for the physical, mental and social well-being of the research subject in all areas connected with the research.

- There is also the principle of doing no harm. The risks incurred by research subjects must be assessed in the light of the benefits that they may derive from the research and the importance of the knowledge likely to be obtained. In any event, the risk to the research subject must always be reduced to a minimum.

To this end, a legislative decree (No. 211) was passed on 24 June 2003, in order to incorporate into Italian law EU Directive 2001/20/EC on the implementation of good clinical practice in the conduct of clinical trials on medicinal products for human use. The decree prohibits private-law companies from offering patients remuneration for agreeing to test a medicinal product (Article 1.5). The fine for doing so ranges from 50 000 to 150 000 euros (Article 22.1).

The decree also prohibits violation of the principle that there must be benefits (Articles 3.1.a, 4.1.d and 5.1.b) and requires that the person concerned (or his or her guardian) be informed of the aims of the trial and the risks and constraints involved (Articles 3.1.b, 4.1.d), that research subjects should not be subjected to physical or psychological duress (Article 3.1.c), that their consent (or that of their guardian) be obtained, on the understanding that it may be withdrawn at any time (Articles 3.1.d, 3.1.e, 4.1.a and 5.1.a), and that subjects be given the contact details of a person with the authority to provide further information (Article 3.1.g). Fines range from 20 000 to 60 000 euros (Article 22.2).

In the case of minors, the decree requires that they be received by someone specialising in dealing with young people, who will explain the nature of the trial to them (Article 4.1.b). In addition, the person in charge of the trial must take into consideration a refusal on the part of a minor (Article 4.1.c) or an adult incapable of giving legal consent (Article 5.1.c). The person in charge may therefore override such a refusal if he or she has obtained the consent of the parents or guardian. It should be noted that there is no penalty for failure to comply with these provisions.

Furthermore, there is no express provision to the effect that the parents (or guardian) of a minor or adult incapable of giving legal consent must receive the contact details of a person with the authority to provide further information. As there is a penalty for failure to comply with this requirement in the case of a person capable of giving legal consent, it is likely that the same will apply by analogy to the legal representatives of a minor.

The clinical trials must have been authorised by the competent ethics committee, failing which a fine of 100 000 to 500 000 euros is incurred (Article 22.5). The ethics committee is an independent body that includes members of the health professions, among others. In its (written) opinion, the ethics committee must assess (Article 6) the relevance and advantages of the clinical trial, the risk/benefit ratio, the protocol, the importance of those in charge of the trial, the biomedical research subjects' files, the quality of the healthcare institution and whether the research subjects (or their legal representatives) have been properly informed. It is also responsible for addressing, and covering in its opinion, such matters as the existence of insurance, any remuneration given to healthy subjects taking part in the trial and the way in which the subjects are contacted in connection with the biomedical research project.

The ethics committee responsible is the one attached to the healthcare institution in which the research is taking place. The ethics committee has thirty days in which to draft an opinion, which it must forward to the Ministry of Health (Article 7). In the absence of an opinion, the biomedical research cannot start, at any rate if it is connected in any way with gene therapy. In any event, biomedical research that modifies the individual's germ-line genetic identity (Article 9.6) is prohibited.

During the biomedical research, the person in charge of the trial may amend the protocol, but is required to inform the Ministry of Health and the ethics committee. It is also specified that the manufacture (or import) of medicinal products for use in biomedical research is subject to the prior approval of the Ministry of Health (Article 13). Inspectors from the Ministry of Health may investigate the nature of the biomedical research undertaken.

In respect of points not covered by the legislative decree (No. 211) on the implementation of good clinical practice in the conduct of clinical trials on medicinal products for human use, biomedical research is governed by Ministry of Health Circular No. 6 of 2 September 2002. This covers, in particular, all research involving the use of medicinal products already on the market with a view to ascertaining their effects.

Such observational clinical trials must also be endorsed by an ethics committee, whose opinion, like all ethics committee opinions on biological research, must be made public through its transmission to the Ministry of Health's Directorate General for the Assessment of Medicinal Products and Pharmacovigilance, National Clinical Trials Observatory (http ://oss-sper-clin.sanita.it).

Furthermore, a decree of 26 April 2002 issued by the Istituto Superiore di Sanità (Health Institute) indicates the tests that must first be carried out to ensure that medicinal products used for biomedical research on human beings are harmless. These guidelines are based on the standards of the European Agency for the Evaluation of Medicinal Products (Emea).

The "Guidelines for ethics committees in Italy" (*Orientamenti per i comitati etici in Italia*) of 13 July 2001, issued by Italy's National Bioethics Committee (Comitato Nazionale per la Bioetica), specify the membership of ethics committees and how they should operate. The key feature is the recommendation that a distinction be drawn in ethics committees between people responsible for monitoring biomedical research and those responsible for trials of new medicinal products.

Although it by no means wants to set up a team of bioethics specialists, Italy's National Bioethics Committee stresses the importance of providing all medical and paramedical staff with training in bioethics. The committee considers it appropriate that those responsible for deciding on the ethics of biomedical research should, by this means, have a sound knowledge of the subject. In addition, by making this distinction, the National Bioethics Committee wanted to avoid the tendency for opinions issued by the members of ethics committees (which usually have an excessive workload) to become "bureaucratic". The suggestion that a Bioethics Committee and a Biomedical Research Commission be set up side by side was not acted on when legislation was introduced.

Like many European countries, Italy has had lengthy and acrimonious bioethical debates about whether or not it should be possible to carry out biomedical research on embryos and stem cells. Italy's National Bioethics Committee has issued three opinions on the subject. The first, dated 22 June 1996, is entitled *Identità et statuto dell'embrione umano* (Identity and status of the human embryo). The second, dated 27 October 2000, is entitled

Sull'impiego terapeutico delle cellule staminali (The therapeutic use of stem cells), while the third, dated 11 April 2003, is entitled *Su ricerche utilizzanti embrioni umani et cellule staminali* (Research using human embryos and stem cells). The third opinion did not elicit a unanimous vote on the issue.

In response to a request from the Minister for Research, the National Bioethics Committee expressed a view (in connection with the European Union's Sixth Framework Programme for Research) as to whether it was ethical to carry out research in Italy using stem cells from human embryos – particularly surplus embryos destined to be destroyed. A majority of the members endorsed the spirit of the first two opinions and rejected any kind of research on embryos or stem cells (except "adult" stem cells and those from the umbilical cord or a miscarriage).

The majority of members of the National Bioethics Committee took the view that "human embryos are complete human beings" and that "the Nice Treaty recognises the dignity of all human beings" and therefore condemned "all public funding in this area which reinforces the view that an embryo is simply a mass of cells". They added that there was "no logical reason" for using stem cells from surplus embryos "except occasionally, for pragmatic reasons" and that "this would, in particular encourage the production of embryos for research purposes".

A minority of members of Italy's National Bioethics Committee, on the other hand, took the view that "taking stem cells from an embryo that is not destined to be implanted in no way reflects a lack of respect for the embryo, but may be considered, if not a contribution from the donor couple, at least as an act of solidarity that enables researchers to devise methods of treating diseases that are difficult to cure". Lastly, a small minority rejected the idea of using stem cells from human embryos except in the case of embryos that were already surplus to requirements, and then only within very restricted limits.

A recent law (passed early in 2004) on medically assisted procreation reflected the opinion of the majority of the members of the National Bioethics Committee. Section 1 states that "in order to facilitate a solution to reproductive problems stemming from human sterility or infertility, it is permissible to resort to medically assisted procreation under the conditions laid down in this law, which safeguards the rights of all the subjects concerned, including the unborn child". The law talks of the "*concepito*", which may be translated as "unborn child" or "child conceived". In any event, under the law the embryo is a person recognised by law.

Section 13 of the law, entitled "Experiments on human embryos", prohibits "all research on human embryos" (*E' vietata qualsiasi sperimentazione su ciascun embrione umano*) in paragraph 1. Paragraph 2 adds that research may be carried out on a human embryo solely for the benefit of "the health and development of the embryo". Paragraph 3 adds that, in all cases, "the production of human embryos for research purposes ... and the selection of embryos for eugenic purposes shall be prohibited" (*Sono, comunque, vietati la produzione di embrioni umani a fini di ricerca ... e ogni forma di selezione a scopo eugenetico degli embrioni*).

This paragraph also prohibits human cloning for reproductive or research purposes (*clonazione dell'embrione sia a fini procreativi sia di ricerca*). Incidentally, the statements by Dr Antinori on his attempts to create clones have made public opinion much more aware of this issue. In an opinion dated 17 October 1997, Italy's National Bioethics Committee had already condemned human cloning. It rejected it again in a motion dated 17 January 2003.

Because it has not been fully ratified by Italy, I have not referred to the Oviedo Convention for the Protection of Human Rights and Dignity of the Human Being with regard to the Application of Biology and Medicine. Despite numerous requests from the National Bioethics Committee, no legislative initiative in this area is in the offing.

With regard to biomedical research on animals, Legislative Decree No. 116 of 27 January 1992 states that biomedical research on animals "may be carried out only to obtain results that cannot be obtained by other scientifically valid and applicable methods that do not involve the use of animals". Furthermore, Law 413/93 allows researchers who have a conscientious objection to so doing to avoid using animals in their biomedical research.

Animals used in biomedical research in Italy, 2000

Purpose of research activity	No.	%
pure research	274 673	30.0
research & development, production and quality control: products and apparatus for medicine and dentistry	518 025	57.0
research & development: products and apparatus for veterinary medicine	14 248	1.6
toxicity tests	64 233	7.0
diagnosis of diseases	25 670	2.8
education	2 835	0.3
other	5 919	0.7

In 2000, 95% of the animals used were mice, rats or other rodents.

Source: Official Gazette, General Series (*Serie Generale*) No. 297- 30.11.2001.

Financially speaking, biomedical research in Italy is not in a very favourable situation. According to information from ISRDS-CNR (Scientific Documentation and Research Institute – National Research Council), total expenditure on biomedical research (by private-law and public-law institutions) in 2001 accounted for 1 540 million euros, or barely more than 0.1% of the gross domestic product. It should be added that 95% of this sum is financed either by the public sector (nearly 48%) or the pharmaceutical industry (nearly 48%). The overall contribution made by funds raised by charities (such as Telethon) accounts for roughly 5% of the total expenditure on biomedical research.

In 2001 public spending on biomedical research amounted to 770 million euros. The main sources of public funding were:

• the Ministry of Health and the Ministry of Research: 370 million euros;

• major research institutions – CNR (National Research Council), ENEA (Organisation for New Technologies, Energy and the Environment): 266 million euros;

• the Health Institute (Istituto Superiore di Sanità): 93 million euros.

In 2001 the Italian pharmaceutical industry spent 769 million euros, or 0.063% of the GDP – less than half the European average (0.190%), and less than Japan (0.153%) and the United States (0.235%).

The lack of investment in biomedical research has caused numerous Italian researchers to go abroad. The 2004 Budget Act has, however, provided for financial assistance to reverse this "brain drain". A survey conducted in 2002 by Censis (Centre for Social Studies and Policies) revealed that the reasons that prompted Italian researchers to go abroad were the higher funding for research (59.6%), financial incentives (56%) and faster career development (50.9%).

Biomedical research in Italy has, however, produced some outstanding achievements. For instance, in October 2002 the MIA (Microscopy and Image Analysis centre) was set up in Monza. It is a European centre of excellence for the analysis of microscope images for biomedical research. The initiative is the result of a partnership between the University of Milan-Bicocca, the Mario Negri Pharmacological Institute and the M. Tettamanti Foundation. In 2003, in southern Italy (the poor relation in regional terms in Italy), a similar initiative (in Ariano Irpino-Campania) made it possible to set up a centre of excellence for genetic research, one of the aims of which is to encourage Italian researchers to return to Italy.

The thirteenth Italian scientific and technological culture week was held in Naples in March 2003. The title of the gathering was "From DNA to the human genome: 50 years of progress in elucidating the mystery of life". The occasion elicited an appeal from Renato Dulbecco (Nobel Prize for medicine), who said: "Italy is second to none, but the lack of research facilities and funds for research has significantly hampered results. There is no shortage of scientific discoveries in Italian laboratories, but it is necessary to make the most of the professional skills that exist and create conditions conducive to the steady progress of Italian biomedical research".

United Kingdom: data protection and confidentiality

by Vivienne Harpwood and Michel Coleman

Medical confidentiality has been recognised as fundamental to codes of medical ethics for centuries,[1] and it is widely regarded as an essential practical component in the successful treatment of patients.[2] In the United Kingdom, the National Health Service (NHS) Code of Practice 2003 states:[3]

> Patients entrust us with and allow us to gather sensitive information relating to their health as part of their seeking treatment. They do so in confidence and they have the legitimate expectation that staff will respect their privacy and act accordingly.

Despite the many commitments to respect for the secrecy and security of identifiable medical information, it is recognised that absolute confidentiality – only the patient and his or her doctor having access to the information – cannot be achieved in the modern world, where patients' records are regularly circulated even outside what are strictly regarded as the medical professions.[4]

It is also accepted that unless information about patients suffering from certain categories of diseases is collected on a large scale, it is impossible to predict trends and other vital developments in the public health arena.[5] The Department of Health acknowledges that public health surveillance must be both reliable and sufficiently comprehensive so that changes can be tracked over time, noting that "if significant levels of patients did opt out of ... surveys, this would call into question their validity, with possible adverse implications over time for public health itself".[6]

In the UK, the desirable balance between confidentiality and the permissible uses of identifiable health information has become confused, mainly as a result of the Data Protection Act 1998 (DPA), the implementation of the Human Rights Act 1998 (HRA) in the year 2002, and professional guidance issued by the General Medical Council in 2000. This chapter examines the status of medical confidentiality in the UK in the context of health surveillance, with particular reference to cancer research and the information collected and held by the cancer registries.

1.
For example, the Hippocratic Oath; The Declaration of the World Medical Association, Geneva, as amended in Sydney 1968 and Stockholm 1994; International Code of Medical Ethics, as amended in Venice 1983; "Duties of a Doctor" General Medical Council, May 2002; British Medical Association October 1999.

2.
See Kennedy I, Grubb A, "Principles of Medical Law", Butterworths, England 1998.

3.
Department of Health, Code of Practice on Confidentiality, DH, London 2003.

4.
Mason JK, McCall Smith RA, Laurie GT, "Law and Medical Ethics", p 240, Butterworths, England 2002.

5.
Adam D, "Data protection law threatens to derail UK epidemiology studies", Nature 411, 31 May 2001.

6.
Department of Health. Human bodies, human choices: the law on human organs and tissue in England and Wales. A consultation report. Department of Health, London 2000.

Information and health surveillance

Health surveillance systems facilitate research on trends in various diseases, in their management and in their impact on the public health. Health surveillance can be described as "the ongoing collection, collation and analysis of data and the prompt dissemination of the resulting information to those who need to know so that action can result".[1] Cancer registries are an integral part of this process.

"Cancer" describes a group of diseases that affect about one in three persons in developed countries, that kill one in four, and which have huge economic and social costs: the importance of cancer in public health is unchallenged. Cancer registries collect and hold long-term identifiable information about the person, the cancer, the treatment and the outcome for all persons diagnosed with cancer among the resident population of the territory they cover. They handle clinical requests for information relating to particular patients, and for counselling healthy persons about genetic risks.

The information they generate is used for public health surveillance and for epidemiological research into the causes, trends and management of disease. This is invaluable in improving the diagnosis and treatment of patients and in the management and control of the disease in the entire community. The UK Department of Health lists a range of cancer registry functions on its website; it also sets out why identifiable data are required, and some of the purposes for which they are used.[2]

Central to the principle that it is acceptable for identifiable information to be divulged by a doctor to others is the notion that the patient's consent will legitimate departures from strict medical confidentiality. This notion is present in ethical guidance,[3] at common law[4] and in legislation.[5] In the UK, the emphasis on this position has increased:[6]

> Requirements for patients to consent to processing of their identifiable healthcare information have increased since the 1990s. The pendulum has swung away from implied consent, the basis of medical research for decades.

1.
"*The Oxford Handbook of Public Health Practice*", ed Pencheon D, Guest C, Melzer D, Muir Gray JA, Oxford, Oxford University Press, 2001.

2.
http://www.doh.gov.uk/cancer/cancer-registries.htm, accessed 6 February 2004.

3.
"Duties of a Doctor", General Medical Council, London, May 2002; British Medical Association, October 1999.

4.
Hunter v Mann [1974] QB 767.

5.
Data Protection Act 1998.

6.
Coleman MP, Evans BG, Barrett G, "Confidentiality and the public interest in medical research – will we ever get it right?" *Clinical Medicine*, Vol 3, No 3, May 2003.

However, as will be seen, in the UK in certain circumstances it is not unlawful for information about patients to be divulged and collected, even without their consent. This limitation of the need for patient consent has prompted some concern. The fears are magnified because in the field of consent to medical treatment and research more generally, UK law is increasingly prepared to maximise patient autonomy, and government policy and the prevailing culture in healthcare demands that patients be "put at the centre".[1] This approach indicates that the UK courts are moving closer to the position in North America[2] and Australia.[3]

The legal framework

The legal framework dealing with medical confidentiality is complex. It derives from contractual obligations, from common law and from statutes. The position at common law was stated in Hunter *v.* Mann[4] as follows :

> The doctor is under a duty not to disclose, without the consent of the patient, information which he, the doctor has gained in his professional capacity.

Exceptions have been developed to this basic proposition over the years, and the common law now recognises that the patient's consent is not necessary if, for example, as a result of a balancing exercise, it would be considered in the overriding public interest to divulge the information,[5] or if the identity of the patients concerned is protected by anonymisation, even if the information is divulged to further the commercial interests of a third party.[6]

The common law and the exceptions that it recognises are relatively straightforward, but the position under the Data Protection Act 1998 (DPA) – which must be read in the light of the Human Rights Act 1998 (HRA) – is not.

The DPA was enacted to bring the UK into compliance with the European directive,[7] which aims to harmonise the law on the processing of personal data in member states. The act is a complex and convoluted piece of legislation, and a full discussion is beyond the scope of this chapter. Suffice it to note that recent confusion about the scope and application of the UK

1.
"*NHS Plan: A Plan for Investment, A Plan for Reform*", London, Department of Health, 2002.

2.
Reibl v Hughes (1980) 114 DLR (3rd) 1.

3.
Rogers v Whitaker (1992) 109 ALR 625 (H.Ct Australia).

4.
[1974] QB 767, 772.

5.
W v Egdell [1990] 1 All ER 835.

6.
R v Department of Health, *ex parte* Source Informatics [2000] 1 All ER 786 CA. The validity of this decision has been questioned – see Beyleveld D, Townend P, *Medical Law International*, ed Longley D and Harpwood V, Vol 6.2, 2004.

7.
Directive 95/46/EC. Protection of individuals with regard to the processing of personal data and on the free movement of such data. European Parliament, 0031-0050.

Data Protection Act – outside the medical context – has led to several deaths, and forced the UK Government to consider an expert review of some of its provisions.[1]

The DPA 1998 updates earlier data protection legislation in the UK. It provides additional protection for medical information, over and above that which is recognised by the common law. It applies only to identifiable living individuals, and covers all information held or processed in any way about data subjects – for our purposes, patients. It prohibits processing unless the data controller is registered under the act, and it creates the post of Information Commissioner to supervise the operation of the law and to guide data users and the public on its operation.

The act places specific duties on data controllers, permits access by individuals to data held about them, and creates specific offences and remedies. There is special protection under the act for personal data, defined as "Any information relating to an identified or identifiable natural person". An identifiable person is "one who can be identified directly or indirectly, in particular by reference to an identification number or to one or more factors specific to his physical, physiological, mental, economic, cultural or social identity".[2]

An important feature of the act lies in its reference to a set of eight data protection principles, within which the law regulates the processing of data. Among the principles of particular relevance for medical data are the following:

- Personal data shall be processed fairly and lawfully and, in particular, shall not be processed unless (a) at least one of the conditions in Schedule 2 is met, and (b) in the case of sensitive personal data [the category to which health records belong], at least one of the conditions in Schedule 3 is met [see below];
- Personal data shall be obtained only for one or more specified and lawful purposes, and shall not be further processed in any manner incompatible with that purpose or those purposes;
- Personal data shall be adequate, relevant and not excessive in relation to the purposes for which they are processed;
- Personal data shall be accurate and, where necessary, kept up to date;
- Personal data processed for any purpose or purposes shall not be kept for longer than is necessary for that purpose or those purposes.

1.
For example, because police officers misunderstood the provisions in the act about data retention, they shredded key information about allegations of sexual assault against a man who was later employed in a school; he was later convicted of murdering two children from the school. In another incident, British Gas incorrectly believed that it was prevented by the act from giving information to social services about an elderly couple whose gas supply was cut off during the winter of 2003: the couple were later found dead from hypothermia.

2.
Data Protection Act 1998, Article 2.

Because health records fall into the category of "sensitive personal data" defined in the act,[1] they are given additional protection under Schedule 3, and can normally be processed only if at least one of the conditions in Schedule 1 is met and at least one of the conditions in Schedule 2 is met. In effect, this means that health data can only be processed either with the explicit consent of the patient or, in special circumstances, without consent. The special circumstances under which explicit consent is not required include when:[2]

The processing is necessary

(a) in order to protect the vital interests of the data subject or another person in a case where –

(i) the consent cannot be given by or on behalf of the data subject, or

(ii) the data controller cannot reasonably be expected to obtain the consent of the data subject, or

(b) in order to protect the vital interests of another person, in a case where consent by or on behalf of the data protection subject has been unreasonably withheld.

The processing is necessary for "medical purposes"[3] and is undertaken by –

(a) a health professional, or

(b) a person who in the circumstances owes a duty of confidentiality which is equivalent to that which would arise if that person were a health professional.

"Medical purposes" is defined as including the purposes of preventative medicine, medical diagnosis, medical research, the provision of care and treatment, and the management of healthcare services.

A need for legislative intervention?

This wide range of exceptions to the requirement for patient consent for processing of health data to be legitimate under the act, in addition to exceptions recognised by common law, would suggest that there is no obstacle to the collection and processing of data for the purposes of health surveillance even without consent, and that cancer registries could safely operate without further legislative or other intervention.

1.
Data Protection Act 1998, Section 2.

2.
Data Protection Act 1998, Schedule 3 (8).

3.
Data Protection Act 1998, Schedule 3 (8).

However, considerable confusion and uncertainty remain as to whether this is in fact the case,[1] and the weight of opinion favoured the need for legislation to clarify the issue. We take issue with that view later in this chapter.

Both the Data Protection Act – which requires processing to be both "fair" and "lawful"[2] – and the common law must comply with the Human Rights Act 1998, and therefore with the ECHR.

According to Department of Health guidance, the principle of "fairness" under the first data protection principle would suggest that before their data are collected or used, patients must be informed of who will have access to the data (that is, the identity of the data controller),[3] and the purposes for which the data are to be processed.[4] They must also be given any further information required to enable the processing to be fair[5] – apparently this means the information that the Department of Health considers is required at common law to support choice as to how their data might be used – and they must be told of their right to impose restrictions upon its use.

For processing to be "lawful", the act requires information to be processed in accordance both with the provisions of the act itself, and with common law.[6] Again, according to the Department of Health, this requires that patients be provided with sufficient information to support their decisions.[7] Since implementation of the Human Rights Act, lawful processing also requires the information to be processed in compliance with the patient's ECHR rights – of which more later in this chapter.

The DPA does include specific exemptions relating to the rights of patients to give consent to the processing of their data,[8] but it does not contain specific exemption from any common-law requirement that patients be given sufficient information about potential uses of their data to support their choices concerning consent to such uses. The Department of Health advice[9] appears to be based on a vague assumption that there is indeed such a common-law requirement, not exempted by the act, to give patients adequate information to enable them to make considered choices concerning the use of their data, and to impose restrictions upon how their data may

1.
See Revill J, "Data law blocks crucial cancer research plans", the *Guardian*, 5 January 2003.

2.
Data Protection Act 1998, Schedule 1.

3.
Data Protection Act 1998, Schedule 1, Part II, (3)(a),(b).

4.
Data Protection Act 1998, Schedule 1, Part II, (3)(c).

5.
Data Protection Act 1998, Schedule 1, Part II, (3)(d).

6.
See "Confidentiality: NHS Code of Practice" para. 26 *et seq.* "Informing Patients", Version 3, 2003.

7.
Ibid.

8.
Data Protection Act 1998, Schedule 3 (8) *supra*

9.
"Confidentiality: NHS Code of Practice," *op. cit.*, para. 26.

be used. However, the cases decided so far at common law relate only to consent to medical treatment: they have not been extended by the courts to consent to the confidential use in research of identifiable information obtained in the course of such treatment.

The most recent advice issued by the General Medical Council (GMC)[1] to doctors on confidentiality,[2] which in the past has been accepted without question, and endorsed by UK courts,[3] states:

> Seeking patients' consent to disclosure is part of good communication between doctors and patients and is an essential part of respect for patients' autonomy. When seeking express consent you should make sure that patients are given enough information on which to base their decision, the reasons for the disclosure and the likely consequences of the disclosure.

Nevertheless, on this precise issue the courts have yet to pronounce. It would seem, therefore, that in its advice on confidentiality, the Department of Health is taking an understandably cautious approach, and on the basis of the general drive towards greater respect for patients' autonomy, includes consent to the use of confidential information within the legal framework that is currently developing in relation to consent to treatment.

Under the common law on consent to treatment, the competent patient must give voluntary and continuing permission to the medical treatment proposed, though in some circumstances a patient may be taken to have given implied consent. It must be emphasised, however, that although the Department of Health's general guidance on consent to medical treatment[4] seeks to promote patient autonomy, the law in this area is as yet by no means fixed. It is even possible to find contradictory statements at the Court of Appeal on the information required for a patient to consent to medical treatment. For example, in Pearce *v.* United Bristol Healthcare Trust,[5] Lord Woolf stated the position as follows:

> If there is a significant risk which would affect the judgement of a reasonable patient, then in the normal course of events it is the responsibility of the doctor to inform the patient of that risk.

1.
The General Medical Council is a statutory body with the duty to regulate the medical profession.

2.
General Medical Council, "Confidentiality: Protecting and Providing Information", London, GMC, 2000.

3.
For example, in X v. Y [1988] 2 All ER 648.

4.
Department of Health, "Reference Guide to Consent", London, DH, 2001.

5.
[1998] 48 BMLR 118.

By contrast, in Burke *v.* Leeds Health Authority,[1] a differently constituted Court of Appeal took the view that what a doctor had to tell a patient, at what point and with what force, were matters for the clinical judgment of the doctor.

It follows from this discussion that legislative intervention in the UK to create further exceptions to the Data Protection Act and the common law may never have been necessary. This view received authoritative support from the responsible cabinet minister in the House of Lords who said, when discussing an amendment to the Freedom of Information Bill,[2]

> the concern ... is ... that the Data Protection Act 1998 does not prevent the medical data of individuals being used for certain medical research purposes, notably, but not solely, in relation to cancer registries [...]. I assure your Lordships that the Act does not have that effect. At present, the 1998 Act allows medical data to be used for any medical research purpose without the need for the consent of individuals. It is not necessary to define the term "medical research", nor to make specific provision for it to include the monitoring of public health, which for these purposes is regarded as medical research.

Additional support comes from the European directive itself, which exempts from its provisions any data

> required for the purposes of preventive medicine, medical diagnosis, the provision of care or treatment or the management of health-care services, and where those data are processed by a health professional subject under national law or rules established by national competent bodies to the obligation of professional secrecy or by another person also subject to an equivalent obligation of secrecy.

Such broad categories of exemption would certainly cover health surveillance and the work of organisations such as the cancer registries, although of course the UK as a member state of the European Union was entitled to draft the legislation more tightly, so as to offer more stringent safeguards for the confidentiality of its citizens than those provided under the European directive. The problem arises when the interpretation of those extra safeguards compromises medical research and public health surveillance, because their consequences have not been adequately taken into account.

1.
[2001], EWCA Civ. 51.

2.
Lord Falconer of Thoroton, on amendment 65A to the Freedom of Information Bill, HL *Hansard*, 14 November 2000, col. 265.

Soon after the Data Protection Act became law, grave concerns were expressed by the scientific community about the perceived threat to the prospects for health research. Numerous letters in scientific journals and the media expressed fears about the future of research,[1] and in particular cancer research, with its emotive connotations for the general public. Attention was drawn in the medical press[2] to the uncertainties of healthcare workers faced with interpreting the DPA. The concerns appear to have been aggravated by the GMC, which issued guidance indicating that, by October 2001, cancer registries had to create new mechanisms for seeking and recording consent for processing their data from every cancer patient about whom they held information.[3]

The House of Commons Select Committee on Science and Technology considered the issue of confidentiality in its inquiry into cancer research. Professor Bruce Ponder argued in his evidence that if it were indeed the case that explicit consent of patients was necessary for the processing of their data, "it would make both cancer registration and almost all cancer epidemiological research effectively impossible".[4] The committee heard substantial evidence on those lines, and recommended that "as a matter of urgency the Advisory Group on Patient Confidentiality should address the concerns posed by the 1998 DPA regarding the registration of cancer".[5] It also recommended that cancer registration should become a statutory requirement, in part to resolve the issue of requiring consent, but the government rejected this proposal.

Support for cancer registries created momentum for clarification of the law, so as to allow cancer and communicable disease registries to conduct their valuable work in a way that was visibly legal. The GMC agreed to suspend its guidance relating to cancer registration until October 2001, to allow time for the Department of Health to find a solution.

Is Section 60 of the Health and Social Care Act 2001 the solution?

The government response to these widespread concerns about confusion surrounding the DPA, the common law and the

1.
For example, the *Guardian*, 12 April 2001.

2.
Strobl J, Cave E, Walley T, "Data Protection legislation: interpretation and barriers to research", *Br Med J* 2000; 321:890-892 in which five NHS Trusts contacted by the researchers took considerable lengths of time to return different answers as to whether patients needed to give explicit consent to the processing of their data; see also Helliwell T, Hinde S, Warren V, "Cancer registries fear collapse" *Br Med J* 2001; 322:730a.

3.
For further reactions see Leslie SJ, Webb DJ, "GMC's guidance may inhibit research" *Br Med J* 2001; 322:1601-1602.

4.
This view was based on advice from one Multi-Centre Research Ethics Committee (MREC) that epidemiological studies were illegal, because cancer registration itself was illegal – see House of Commons Science and Technology Committee, Sixth Report 2001, para. 101.

5.
Ibid para. 112.

GMC's guidance, was to insert a special provision into a bill focused primarily on other healthcare matters. Section 60 of the Health and Social Care Act 2001 (HSCA), as it became, provided a mechanism for certain classes of identifiable health data to be legally processed without the patient's consent, thus modifying the common law :[1]

> The Secretary of State may by regulations make such provision for and in connection with requiring or regulating the processing of prescribed patient information for medical purposes as he considers necessary or expedient –
>
> (a) in the interests of improving patient care
>
> (b) in the public interest
>
> "Medical purposes" are defined here as any of the following :
>
> (a) preventative medicine, medical diagnosis, medical research, the provision of care and treatment and the management of health and social services, and
>
> (b) informing individuals about their physical or mental health or condition, the diagnosis of their condition or their care or treatment.[2]

It also created a new statutory body, the Patient Information and Advisory Group (PIAG), under Section 61 of the Act, to provide safeguards to control the operation of Section 60 and regulations made under it.

But the Government views Section 60 support for the confidential use of identifiable information for research as "transitional", until such time as anonymisation obviates the need for patient identifiers, or adequate measures for obtaining patient consent can be instituted. The Health Minister expressed it very simply in Parliamentary debate :[3] "as soon as we can, we will take away that support".

Even Section 60 did not receive an unqualified welcome in its passage through Parliament, and the wide discretion it gave to the Secretary of State was criticised in several quarters.[4] The introduction of such far-reaching powers gave rise to concerns about compliance with the European Convention on Human Rights.

1.
Health and Social Care Act, Section 60 (1).

2.
Health and Social Care Act, Section 60 (1)(10).

3.
Denham, J. Health and Social Care Bill (Standing Committee E). HC *Hansard* 2001 ; 8 February : c 480.

4.
For example, Michael Wilkes of the British Medical Association said, "Informed patient consent is a fundamental tenet of good medical practice. I am unhappy about giving the Secretary of State such wide powers."

The Human Rights Act requires all public bodies – including government bodies and ministers, local authorities, courts, NHS Trusts and disease registries – to act in accordance with the rights identified in the European Convention on Human Rights. When interpreting legislation, and when human rights issues are raised before them, UK courts are required to take into account the jurisprudence of the European Court of Human Rights.[1] If a higher court in the UK considers that any statutory provision infringes a Convention right, it may issue a declaration of incompatibility for Parliament to deal with as appropriate.

In the context of confidentiality, Article 8 is the relevant article of the European Convention on Human Rights, which deals with the right to privacy and family life.[2] The European Court of Human Rights has ruled that medical records are covered by this article.[3] Article 8 rights are not absolute, and derogations are permitted in certain circumstances stated in Article 8 (2) :

> There shall be no interference by a public authority with the exercise of this right except such as is in accordance with law and is necessary in a democratic society in the interests of national security, public safety or the economic well-being of the country, for the prevention of disorder or crime, for the protection of health or morals, or the protection of the rights and freedoms of others.

Although measures taken by member states for the protection of health, such as those legitimated by the DPA and by Section 60 of the HSCA and regulations made under it, are clearly covered by the permitted exceptions to Article 8, a ruling of the European Court of Human Rights[4] indicates that :

> Any state measures compelling communications or disclosure of such [confidential] information without the consent of the patient call for the most careful scrutiny.

In particular, in the spirit of Article 8 (2), any departures from the rights in Article 8, including those specified in the UK legislation and regulations considered here, must pursue a legitimate aim; must be necessary in a democratic society, and must be proportionate to the end to be achieved by them. In other words, the means by which the aim is to be secured must not be excessive, taking into account the political, cultural and

1.
Human Rights Act 1998, Section 2(1)(a) to (c); Section 3(1).

2.
European Convention on Human Rights, Article 8(1) "Everyone has the right to respect for his privacy and family life, his home and his correspondence."

3.
Charre (nee Jullien) v France ; (1998) 25 EHRR 371.

4.
Z v Finland (1998) 25 EHRR 371.

social traditions of the UK, for which a margin of appreciation is permitted.[1] A fair balance must be struck between the demands of the interests of the community and the need to protect the rights of the individual.[2]

The concept of proportionality requires that appropriate safeguards be established by member states when enacting measures that derogate from Convention rights. The legislation under discussion here does contain such safeguards, though in the case of the DPA and the common law in the UK, these are drafted in such a way as to confuse rather than clarify.

The safeguards introduced by the Health and Social Care Act 2001 include the setting-up of the Patient Information Advisory Group, the remit of which is to comment upon any regulations proposed under Section 60 by the Secretary of State for Health. The safeguards also include the caveats that regulations be reviewed each year,[3] that relevant interest groups be consulted,[4] and that regulations may not provide for the processing of confidential patient information for any purpose if it would be reasonably practicable to achieve that purpose in some other way having regard to the cost and technology available.[5] This means that if PIAG's annual review concludes that technical progress since the previous year makes it unnecessary to hold identifiable data without consent for a particular purpose, the relevant regulation must be revoked and the power to use data in that way withdrawn.

Despite the inclusion of such safeguards, numerous objections were raised to Section 60. Earl Howe commented in the House of Lords :[6]

> The dignity and privacy of the individual is being subordinated to the administrative convenience of the NHS .... That extends to giving the Secretary of State power to instruct any health professional to divulge individual patient data even against the express wishes of the patient.

Baroness Finlay countered with the point that research is crucial in the long-term interests of the community, and that cancer registries play a vital part in providing essential data for research.

1.
See Lawless v Ireland [1961] 1 EHRR 15 ; also N. Lavender, "The Problem of the Margin of Appreciation" [1997] EHRLR 380.

2.
Soering v United Kingdom (1989) 11 EHRR 439.

3.
Health and Social Care Act 2001, Section 60 (4)(a)(b).

4.
Health and Social Care Act, Section 60 (7).

5.
Health and Social Care Act, Section 60 (3).

6.
HL, *Hansard*, 21 May 2002.

The impact of the regulations

Regulations made by the Secretary of State under Section 60 came into force in June 2002. Regulation 2 exempts data collection by the cancer registries from the common-law requirement to obtain the patient's consent. The practical consequences of the new legislative safeguards on cancer registries, in terms of the additional workload required to prepare documentation in order to gain PIAG approval for their health surveillance and research activities, and that involved in anonymisation of patients, cannot be underestimated. Although the regulations provided new legal protection for data collection activities by cancer registries, continuing reluctance by some data providers forced the Department of Health to remind all NHS Chief Executives in September 2002: "All organisations are asked to continue submitting information to cancer registries and can be assured that it is legitimate to do so."[1]

It remains to be seen whether the present legislative safeguards are adequate, since no court has yet pronounced on them. The Department of Health advice is that the guidance should be obtained from the data protection officer or Caldicott Guardians[2] and that:

> It is good practice to inform patients where section 60 has been used to set aside common law confidentiality requirements and an exemption to the fair processing requirements of the DPA 1998 has been applied, but there is no requirement to do so if this would require disproportionate effort.[3]

Continuing questions about the legal implications of data processing also create practical difficulties for the registries. The present position raises several critical questions that require clarification:

- For how long should data be retained?

- Is the enforcement of the regulations practical?

- Is the level at which penalties are set for infringement of the regulations adequate?

- Does the requirement for annual review of the need for regulatory approval to process patient information create an unreasonable burden on registries?

1.
Department of Health. Section 60 support for cancer registries. Chief Executive Bulletin 2002;20-26 September: Item 5. http://www.doh.gov.uk/cebulletin26september.htm#5

2.
Following the recommendations of the Caldicott Committee in 1997, NHS organisations each have a Caldicott Guardian with responsibility for patient confidentiality.

3.
Department of Health guidance, *op cit*, para. 25.

- What is the position of researchers outside the NHS who may not be properly regarded as "health professionals" within the safeguard in regulation 7(2)?

- What is the legality of anonymisation and pseudonymisation carried out by registry staff?

- Are the research ethics committees that are required to approve proposed research, using data processed by cancer registries, adequately equipped to answer questions on the complex legal issues that arise in this context?

- Do the interests of research justify a major departure from patient autonomy?

- In human rights terms, do utilitarian arguments centred on the "public good" outweigh the right of the individual patient to confidentiality?

- Is there a possibility of a challenge to a registry under human rights legislation by a patient whose data are processed without explicit consent having been obtained?

- Will patients make requests under Section 7 of the DPA for details about how their data have been processed and who has had access to them?

A significant legal concern for cancer registries is that they are now frequently asked to advise geneticists about the cancer-risk status of individuals. In Wales alone, it has been reported[1] that the number of requests from geneticists to the cancer registry has risen from around 50 in 1998 to about 700 in 2003. Section 60(5) of the Health and Social Care Act 2001 states that regulations may not make provision for requiring the processing of confidential patient information solely or principally for the purpose of determining the care and treatment to be given to particular individuals.

However, it appears that Regulation 2(1)(e) does precisely that. It permits the provision of information about:

> individuals who have suffered from a particular disease or condition where – (i) that information supports an analysis of the risk of developing that disease or condition; and (ii) it is required for the counselling and support of a person who is concerned about the risk of developing that disease or condition.

1.
UK Association of Cancer Registries, 10th Annual Conference Proceedings, 8 November 2003.

It could be argued that this regulation is *ultra vires* (beyond the powers) given to the Secretary of State by Section 60. This requires urgent review and clarification in the interests both of patients and of those responsible for the operation of cancer registries, in the light of human rights considerations. Detailed discussion of this legal problem is beyond the scope of this chapter. A consultation on Section 60 is currently in progress at the behest of the UK Government.

The new statutory provisions in the UK for confidential handling of identifiable health data, for specified purposes and without consent of the data subject, were intended only as an "interim measure", but appear already to have become entrenched. There is doubt over whether such legislation was necessary. A comprehensive review is urgently required of the Data Protection Act 1998 and of Section 60 of the Health and Social Care Act 2001 and the regulations made under it.

The UK public deserves to be better informed[1] about the way in which their health data are processed, not only for cancer information but also in the broader arena of the use of electronic records and modern communication systems in the NHS. Wide consultation and public debate will be required on the appropriate balance to be struck between personal autonomy and the confidential handling of identifiable health information without consent in the public interest.

Consideration should be given to establishing a new statutory framework within which cancer and other disease registries can lawfully process patients' information without their consent, subject to stringent but workable safeguards that take into account both the present human rights framework and the culture of increasing autonomy for patients. The temporary nature of Section 60 support for data collection and processing, and the need for annual review of the regulations, do not provide a stable climate in which to plan the development of disease registration or medical research involving the confidential use of identifiable data for research in the public interest.

Confusion and mistrust created by the current legislative and regulatory position have had a significant adverse impact on medical research. We believe that only primary legislation will

1.
See Coleman et al., *op. cit.*, p. 226.

provide a sufficiently durable equilibrium between personal autonomy, human rights, public health surveillance and medical research – one that will command widespread public support in the twenty-first century.

The national cancer registration system (Extracts from Department of Health website)

Cancer registries are essential to the implementation of the NHS Cancer Plan, which aims to improve the quality of care and survival for cancer patients. Reliable population-based information on cancer incidence, prevalence and survival rates is needed and cancer registries are the only available source.

Uses of cancer registries

Cancer registries undertake a range of public health surveillance and health protection functions. The main functions of cancer registries are:

- monitoring trends in cancer incidence, prevalence and survival over time and between different areas and social groups;

- evaluating the effectiveness of cancer prevention and screening programmes; for example, population-based data are required to monitor the effectiveness of the existing national screening programmes for breast and cervical cancer and to inform the design of new programmes, such as screening for colo-rectal and ovarian cancer;

- evaluating the quality and outcomes of cancer care by providing comparative data about treatment patterns and outcomes;

- evaluating the impact of environmental and social factors on cancer risk. For example, cancer registry data are used to investigate possible cancer risks in relation to power lines, landfill sites and mobile phones. Cancer registry data are also used to investigate differences in cancer incidence, survival and access to treatment between social groups and thus contribute to programmes aimed at reducing inequalities in health outcomes;

- supporting investigations into the causes of cancer;

- providing information in support of cancer genetic counselling services for individuals and families at higher risk of developing cancer.

To fulfil these objectives, cancer registries need to collate data on individual patients from multiple sources and over long periods. These sources include district general hospitals, cancer centres, hospices, private hospitals, cancer screening programmes, other cancer registers, primary care, nursing homes and death certificates. Data are frequently collected from several sources within an individual institution (for example, pathology departments, medical records and radiotherapy databases).

Registries are asked to link their information with information from other NHS services, notably to support the evaluation of the effectiveness of the national breast and cervical cancer screening programmes. Ensuring such linkages are accurate again requires checking of personal identifiers.

Registries are frequently asked by cancer genetic counselling clinics to confirm cancer diagnoses in family members. The individuals concerned must first provide informed consent to the release of their information for this purpose, but in order to provide such information registries need to have included these names on their databases.

Registries also supply names of patients to bona fide researchers for detailed research projects investigating causes of (or outcomes from) specific cancers. Such studies must first be approved by the appropriate multi-centre (or local) research ethics committees. If patients are approached to provide further information (for example, regarding occupation or lifestyle), this approach will always be via the patient's general practitioner or hospital consultant, and participation will depend on the patient's fully informed consent.

Geographical studies (such as studies of cancer risk in people living near landfill sites) can be undertaken only if a full postcode is available. A postcode is also required to convert historic data to existing boundaries of, for example, regions, health authorities and primary care organisations.

Data release

All transfers of data to and from cancer registries are strictly controlled and there have been no breaches of confidentiality in the more than forty-year history of the national cancer registration system. Identifiable data releases to third parties are restricted according to strict data release policies. Publications from cancer registries only present aggregated data; they never identify individual patients.

What do we know as a result of cancer registration?

This is best summarised by listing some examples of what we know as a result of the work of cancer registries and what we will not know in future if cancer registration data become unreliable.

What we know as a result of information obtained from cancer registration:

- mesothelioma is caused by exposure to asbestos;
- skin melanoma rates have been increasing year on year;
- lymphoma and oral cancer rates are higher in ethnic minorities;
- there is wide variation in how cancer is treated around the country;
- cancer survival for patients living in poor areas is lower than for those living in rich areas;
- cancer survival is lower in the UK than in Europe for most cancers; and
- cancer survival in children has improved dramatically over the last thirty years.

What will we not know in future if cancer registration becomes unreliable?

We may not discover:

- how many cancers occur each year, and which are the most common;
- if cancer occurrence rates in the UK are higher or lower than in other countries;
- if cancer survival rates in the UK have caught up with other European countries;
- if inequalities in cancer treatment or survival between rich and poor have been abolished;
- if cancer screening programmes are effective;
- if people living near landfill sites or power lines have an increased cancer risk;
- whether some late deaths in childhood-cancer survivors are related to earlier treatments;
- if the risk of developing certain cancers is elevated in some occupational groups.

Europe and
biomedical research

European law and biomedical research

by Pēteris Zilgalvis[1]

The Council of Europe was the first international organisation to prepare a binding international agreement addressing the new biomedical technologies. The Convention on Human Rights and Biomedicine was opened for signature on 4 April 1997 in Oviedo, Spain and thirty-one countries[2] have signed to date. Eighteen member states have ratified and the convention has come into force for these countries.[3] However, the convention was not the Council of Europe's first foray into bioethics or biomedical research, but was preceded by a number of recommendations in this field and a great deal of debate at intergovernmental level. The convention has since been followed by an Additional Protocol on Biomedical Research further extending the Council's contribution in this domain.

In this chapter, the work preceding the convention, the convention itself, and the continuing work building on the achievements of the convention will be discussed, along with relevant contributions to the field made by the European Union institutions.

It is important to note that the instruments of the Council of Europe and the EU do not cover identical fields. The Council of Europe is quite distinct from the twenty-five nation European Union, though all the EU member countries are also members of the Council of Europe. The obvious difference between the two institutions is the Council of Europe's broader membership of forty-five European countries, stretching from one end of Europe to the other. The Secretary General of the Council of Europe, Mr Walter Schwimmer, emphasised in his speech on the role and place of the Council of Europe in the context of the enlargement of the European Union, at the Paris Press Club on 3 July 2001, that the problems of bioethics are not limited to just one part of Europe.[4]

At the same time, it is necessary to acknowledge that the contributions of European Union institutions are also quite pertinent.

1.
The views expressed are personal and do not necessarily reflect any official position of the Council of Europe.

2.
As of 1 September 2004: Bulgaria, Croatia, Cyprus, Czech Republic, Denmark, Estonia, Finland, France, Georgia, Greece, Hungary, Iceland, Italy, Latvia, Lithuania, Luxembourg, Moldova, Netherlands, Norway, Poland, Portugal, Romania, San Marino, Slovak Republic, Slovenia, Spain, Sweden, Switzerland, The Former Yugoslav Republic of Macedonia, Turkey, and Ukraine.

3.
Ratifications as of 1 September 2004: Bulgaria, Croatia, Cyprus, Czech Republic, Denmark, Estonia, Georgia, Greece, Hungary, Lithuania, Portugal, Romania, San Marino, Slovak Republic, Slovenia Spain, and Turkey.

4.
http://test.press.coe.int/Discours/Wspressclub-e.htm accessed on 9 July 2001.

The Community is empowered to act in this field on the basis of EC Treaty Articles 152 (public health), Articles 163 to 173 (research, funding of the research framework programme) and Article 95 (the internal market). Additionally, Article 49 of the Treaty of European Union (TEU) states that candidate countries must respect human rights and fundamental freedoms in order to join the European Union. Direct reference is made in Article 6 of the TEU to the European Convention on Human Rights.

Not least, as a major funding source for research in Europe through its Sixth Framework Programme, the European Commission has direct influence on what is considered ethically acceptable for researchers. In the field of pharmaceutical research, Directive 2001/20/EC of the European Parliament and of the Council on the approximation of the laws, regulations and administrative provisions of the member states relating to the implementation of good clinical practice in the conduct of clinical trials on medicinal products for human use was adopted on 4 April 2001.

It is important to note that while the Council of Europe's Convention on Human Rights and Biomedicine covers all types of biomedical research on human beings, the aforementioned EU directive on good clinical practice deals only with pharmaceutical research. In regard to medicinal products, it has been stated that the "European Community has a clearly established legal competency. The legal basis for Community action is the principle of the free movement of goods in the EU embodied in Article 3 of the Treaty on European Union."[1]

Another interesting development in the European Union is the Charter of Fundamental Rights. Its Article 3, paragraph 2 addresses the fields of medicine and biology directly, requiring free and informed consent and prohibiting eugenic practices, reproductive cloning and the making of the human body and its parts as such into a source of financial gain. It is noteworthy that the explanation of this article found in document Convent 49[2] states that " the principles of Article 3 are already included in the Convention on Human Rights and Biomedicine, adopted by the Council of Europe." It also prohibits any discrimination based on genetic features in its Article 21 (non-discrimination).

1.
Laurence Cordier, "Is there a European ethical framework for clinical research?" *International Journal of Pharmaceutical Medicine*, 1997;, 11 :137-140.

2.
The explanations in document Convent 49 were prepared by the Praesidium of the Convention that prepared the Charter. However, it is stated therein that they have no legal value and are simply intended to clarify the provisions of the Charter.

The legal status of this instrument, though influential, is not entirely clear at the moment. The charter could be described as having a declaratory nature in the present scheme of things in the European Union. However, Mr Romano Prodi, the then President of the European Commission, did state that "In the eyes of the European Commission, by proclaiming the Charter of Fundamental Rights, the European Union institutions have committed themselves to respecting the Charter in everything they do and in every policy they promote".[1] Furthermore, it has been proposed that any future EU constitution could include the Charter of Fundamental Rights.[2] Other commentators have stated that, in any case, it is the most modern international instrument addressing human rights and will be a very influential source for legal argumentation.[3]

In the field of general data protection, international regulation also exists, for example the Council of Europe's Convention for the Protection of Individuals with regard to Automatic Processing of Personal Data, Council of Europe Recommendation (97)5 on the protection of medical data and the directives of the EU related to data protection, EU Directive 95/46/EC on the protection of individuals with regard to the processing of personal data and on the free movement of such data and Directive 96/9 on the legal protection of databases. However, this area is of sufficient complexity that it warrants a detailed discussion of its own.

In the sphere of patenting and biotechnology, the relevant EU legislation is Directive 98/44/CE on the legal protection of biotechnological inventions, which was adopted on 6 July 1998 after protracted discussions in the European Parliament and in the member states of the EU. The goal of the directive is to harmonise national regulations and jurisprudence in EU member states on patenting in the sphere of biotechnology, based on the premise that without the safeguard provided by patents, industry would not be ready to invest in research and development in Europe.

The legality of the directive was challenged by the Netherlands, with Italy and Norway intervening in opposition. The European Court of Justice dismissed the Dutch action seeking annulment of the Community directive on 9 October 2001. Its

1.
www.europarl.eu.int/ charter/default_en.htm accessed on 9 July 2001.

2.
Assemblée Nationale, Délégation pour l'Union Européenne, Compte Rendu No. 149, Réunion du 19 juin 2001, audition de M. Jacques Delors, www.assemblee-nationale.fr/europe, accessed on 9 July 2001.

3.
Egils Levits, "Cilvēktiesī bas Eiropas Savienības tiesību sistēmā, *Likums un Tiesības*, Vol. 2, No. 11 (15), November 2000: 335.

judgment in Case C-377/98 took the view that Directive 98/44/CE frames patent law in a sufficiently rigorous way so as to ensure that human dignity is safeguarded and that the human body remains unpatentable.

Article 6.2.c of the aforementioned Directive 98/44/CE is also of relevance to the use of embryonic stem cells. It states that uses of embryos for industrial or commercial purposes are unpatentable if their commercial exploitation is contrary to public order or morality.

The background of the Convention on Human Rights and Biomedicine

The Council of Europe's Parliamentary Assembly adopted Recommendation 1100 on the use of human embryos and foetuses in scientific research in 1989. This was their first recommendation specifically addressing biomedical research, though recommendations and resolutions in the field of bioethics had been adopted by the Assembly as long ago as 1976, and some of them, for example, Recommendation 934 (1982) on genetic engineering, were also relevant to aspects of biomedical research.

The Committee of Ministers had also adopted resolutions and recommendations dealing with bioethical issues as long ago as 1978, but the first recommendation specifically addressing biomedical research was Recommendation (90)3 concerning medical research on human beings. This recommendation set out some of the fundamental principles for biomedical research on human beings, which were later further developed in the Convention on Human Rights and Biomedicine and its Additional Protocol on Biomedical Research. We find there the principles of the requirement of free, informed, express and specific consent, and the protection of vulnerable persons and legally incapacitated persons in this recommendation.

Returning to the Council of Europe's Convention on Human Rights and Biomedicine: its roots can be traced to the 17th Conference of the European Ministers of Justice (Istanbul, Turkey, 5-7 June 1990), who adopted Resolution No. 3 on bioethics. This recommended that the Committee of Ministers examine the possibility of preparing a framework convention "setting out common general standards for the protection of the human person[1] in the context of the development of the biomedical sciences." The resolution was based on a proposal by Ms Catherine Lalumiere, Secretary General of the Council of Europe at that time.

The Parliamentary Assembly of the Council of Europe recommended in June 1991, in its Recommendation 1160, that the Committee of Ministers "envisage a framework convention comprising a main text with general principles and additional protocols on specific aspects." The support for the proposal continued to grow when in September 1991, the Committee of Ministers instructed the Ad hoc Committee on Bioethics (CAHBI) to prepare a framework convention setting out common general standards for the protection of the human person in the context of the biomedical sciences, and alluded to protocols to this convention on organ transplants and the use of substances of human origin, and on biomedical research.

In July 1994, a first version of the draft convention was opened for public consultation and was submitted to the Parliamentary Assembly for an opinion.[2] The Steering Committee on Bioethics (CDBI), which had replaced the CAHBI, took this opinion and others into account in preparing a final draft. The CDBI confirmed this draft on 7 June 1996 and submitted it to the Parliamentary Assembly for an opinion.[3] The Committee of Ministers adopted the convention on 19 November 1996.[4]

The convention and its protocols are a "system" that can respond to new (and sometimes threatening) developments in biomedicine. An example is the preparation of the additional protocol prohibiting human cloning after the news of Dolly the sheep's birth came out. Another example is Article 29 of the Additional Protocol on Biomedical Research addressing research in non-member states, which was developed in response to allegations of exploitation of research subjects from the developing world, and central and eastern Europe, by researchers from western Europe.

1.
It is interesting to note that the term "human person", used in a number of documents proposing the preparation of the convention, does not appear in the convention itself. The convention uses "everyone", "human being" or "person".

2.
Opinion No. 184 of 2 February 1995, Doc. 7210.

3.
Opinion No. 198 of 26 September 1996, Doc. 7622.

4.
Germany and Belgium requested that their abstention be recorded when the Committee of Ministers voted on the adoption of the convention and the authorisation of publication of the explanatory report.

For the first time, the convention seeks to establish a common, minimum level of protection in the field of biomedicine, and more specifically in the domain of biomedical research, throughout Europe. In the case of biomedical research, a balance also needed to be found between the freedom of research, which brings many benefits to individuals suffering from diseases, and the regulation of research to protect the same or different individuals. Fears may arise that if a type of research is prohibited in a single country, it will fall behind in the progress of its research and become dependent on work done elsewhere.[1] The development of European standards helps to alleviate such worries.

The convention gives precedence to the human being over the sole interest of science or society. The aim of the convention is to protect human rights and dignity and all its articles must be interpreted in this light. The main focus of the convention with regard to biomedical research is specifically this human rights aspect, unlike other legal instruments in the field, which may concentrate, for example, to a large extent on the economic and public health aspects of making new medicines available more quickly. The interests of society and science are not neglected, however, and come immediately after those of the individual. On this basis, it establishes that consent is obligatory for any medical treatment or research and recognises the right of all individuals to have access to information concerning their health. The text also sets out safeguards protecting anyone, of any age, who is unable to give consent.

The term "human rights" as used in the title and text of the convention refers to the principles found in the European Convention for the Protection of Human Rights and Fundamental Freedoms of 4 November 1950, which guarantees the protection of such rights. The Convention on Human Rights and Biomedicine not only shares the same underlying approach, plus many ethical principles and legal concepts, but also elaborates on some of the principles found in the earlier Convention.

Research under the convention

Requirements for any research to be undertaken on human beings are set out in the convention's chapter on Scientific

1.
Bernard Charles et Alain Claeys, "Réviser les lois bioéthiques : quel encadrement pour une recherche et des pratiques médicales maîtrisées ?", Les documents d'information de l'Assemblée Nationale, No. 3208, 2001, pp. 27-8.

Research specifically, and in other chapters. The convention and its Additional Protocol on Biomedical Research apply to all biomedical research involving interventions on human beings.[1] The general rule for scientific research is set out in Article 15. It states that scientific research in biomedicine shall be carried out freely,[2] subject to the provisions of the convention and the other legal provisions ensuring the protection of the human being.

The framework principles for all research on human beings are further enunciated in this relevant chapter. Those principles are: there must be no alternative of comparable effectiveness to research on humans; risks to be incurred by the person shall not be disproportionate to the potential benefits of the research; the necessary precondition of approval of the research project by the competent body after a multidisciplinary review of its ethical acceptability; there must be informed, express, specific and documented consent.

There are, of course, situations where persons are not able to give their consent themselves. They may be small children or be suffering from senile dementia, to give two examples. The chapter foresees protection in the context of biomedical research for all such persons not able to consent. The aforementioned principles listed in the previous paragraph all apply to these persons as well, with the obvious exception of the proviso on the person's consent. The participation of such a person in a research project may be authorised (specifically and in writing) by a parent, guardian or other representative or body provided for by law, subject to the fulfilment of stringent protective conditions. The results of the research must have the potential to produce real and direct benefits to his or her health; it must be the case that research of comparable effectiveness cannot be carried out on individuals capable of giving consent; and there must be no objection on the part of this person.

Article 17 also provides, exceptionally and under the protective conditions prescribed by law, that research which does not have the potential to produce results of direct benefit to the health of a person not able to consent to research may be carried out if stringent conditions are fulfilled. In addition to the

1.
The protocol states in its Article 2, that for the purposes of the protocol, the term "intervention" includes a physical intervention and any other intervention in so far as it involves a risk to the psychological health of the person concerned.

2.
The freedom of scientific research is a constitutionally protected right in some of the member states; see, for example, Article 20 of the Swiss Constitution.

aforementioned requirements for research on persons not able to consent, this research must have the aim of contributing, through significant improvement to the scientific understanding of the individual's condition, disease or disorder, to the ultimate attainment of results capable of conferring benefit to the person concerned or to other persons in the same age category or afflicted with the same disease or disorder or having the same condition. Finally, the research must entail only minimal risk and minimal burden for the individual concerned.

Moving to the following chapter, Article 18 of the convention states that where the law allows research on embryos *in vitro*, it shall ensure adequate protection of the embryo and it stipulates that the creation of human embryos for research purposes is prohibited. This does not mean that research on supernumerary embryos created for fertilisation purposes is prohibited.

Financial gain from the human body and its parts, as such, is prohibited by Article 21 of the convention.[1] The issue of financial gain arising from the human body or its parts is addressed specifically in the context of biomedical research in the Additional Protocol on Biomedical Research and in the draft instrument addressing research on archived human biological materials.

A provision that is of particular relevance to research on biological materials is Article 22 of the convention, covering disposal of a removed part of the human body. It states that when any part of a human body is removed during the course of an intervention, it may be stored and used for a purpose other than that for which it was removed, but only if this is undertaken in conformity with appropriate information and consent procedures.

1.
The explanatory report notes that the question of patents was not considered in connection with this provision; accordingly it was not intended to apply to the issue of the patentability of biotechnological inventions.

Additional Protocol on Biomedical Research

The convention has been supplemented by an Additional Protocol on Biomedical Research. The Protocol on Biomedical Research was approved by the CDBI in June 2003 and adopted by the Council of Europe's Committee of Ministers in June 2004. In addition to this protocol covering interventions on human beings, the CDBI is preparing an instrument on research on archived biological materials, which will address biomedical research on the personal data included in a research project on archived biological materials.[1]

The full range of biomedical research activities involving any kind of intervention on human beings are covered by the protocol. For the purposes of the protocol, the term "intervention" includes both physical interventions and any other intervention posing a psychological risk to the person concerned.

It is worthwhile noting that both protocols, like the convention, will apply to both privately funded and publicly funded research. This departs from the approach taken by the United States, which has often regulated only federally funded research, though there are exceptions (research coming under the authority of the Food and Drugs Administration, for instance).

In addition to reasserting the principles concerning research found in the convention, the protocol further elaborates on them and addresses related issues in detail. The protocol's Article 7 requires that research be undertaken only if the research project has been approved by the competent body in conformity with national law, after independent examination of its scientific merit, and a multidisciplinary review of its ethical acceptability by an ethics committee.

Special attention is being paid in the Council of Europe to the fulfilment of the requirement for a multidisciplinary review of the ethical acceptability of biomedical research; undertaking a programme of co-operation in 1997-2004 with its member countries in central and eastern Europe, called Debra. The independence of ethics committees is paramount. As Senator Claude Huriet, who has served as a rapporteur for a Debra meeting in Vilnius, Lithuania, stated in the French Senate report on the Protection of Persons Undergoing Biomedical Research, the independence of the committees is the foundation of their credibility and legitimacy.[2]

Chapter III of the protocol addresses the question of ethics committees and opens with Article 9 on independent examination by an ethics committee. It requires that research projects be submitted to independent examination in each state in

1.
A preliminary, public version of this text is available at:
http://www.coe.int/T/E/Legal_affairs/Legal_cooperation/Bioethics/

2.
Claude Huriet, "La protection des personnes se prêtant à des recherches biomédicales. La rôle des comités : un bilan et des propositions", Les Rapports du Sénat, No. 267, 2000-2001, p. 15.

which any research activity is to take place. This includes states from which research subjects are to be recruited for research physically carried out in another country. Best practice is to also submit research projects to an ethics committee in every research location within countries. All research projects within the scope of this protocol must be submitted for review.

A positive assessment by the ethics committee is not required, since the role of such bodies or committees in many countries may be solely advisory. The conclusion of this assessment may have legal force in some jurisdictions, while in others it serves to advise the competent body (for example, a regulatory authority) that will make a binding decision on whether the research project can commence. The article sets out the purpose of the multidisciplinary examination after the precondition of scientific quality has been met. It also states that the assessment shall draw on an appropriate range of expertise and experience adequately reflecting professional and lay views.

Chapter IV addresses consent and information. Consent to participation in biomedical research is addressed by Article 14. As noted above, informed consent is a fundamental principle of the convention, with regard to medical or research interventions. Consent can be freely withdrawn by the person at any stage of the research. If the person in question is not able to give consent, then Chapter V (Protection of persons not able to consent to research) applies.

Protection of private life and confidentiality

Chapter VIII of the protocol deals with confidentiality and the right to information. It provides for the confidentiality of any information of a personal nature obtained during biomedical research, the accessibility to research participants of information collected on their health, the availability of research results, the duty of care to research participants concerning information of relevance to their current or future health, and protection of information related to the research.

Research outside Council of Europe member states

Article 29 of chapter VIII addresses research in states that are not parties to the protocol. The background to the article is that it seeks to prevent the possibility of exploitation of potential research subjects in the developing world by scientists trying to undertake research that would be prohibited in their own countries and by the protocol. It sets out the requirement that sponsors and researchers within the jurisdiction of a party to the protocol who plan research in a state not party to the protocol shall ensure that, without prejudice to the conditions applicable in that state, the research project complies with the principles on which the provisions of the protocol are based. Therefore, they cannot go "ethical shopping" and search for jurisdictions with low or non-existing standards for the protection of the human being in research.

In conclusion, the Council of Europe has made a substantial contribution to the protection of human rights in the field of biomedical research. Its activities continue with the twin aims of developing norms on specific aspects of this domain and of ensuring the application of existing norms.

The Council further seeks to stimulate public debate on these subjects in the spirit of pluralism and democracy. It is recognised that there is a need for continued international co-operation in this field to further clarify and strengthen the protection of human rights and dignity in the context of biomedical research on the global level, and the Council of Europe seeks to actively collaborate with its partners internationally and nationally to achieve these goals.

Appendices

Appendix I – Selected websites

American Journal of Bioethics, http ://www.bioethics.net/
Features include Bioethics for Beginners, Bioethics and Genetics,
Bioethics in Other Journals and a Bioethics Forum.

American Society of Bioethics and Humanities,
http ://www.asbh.org/
The American Society for Bioethics and Humanities (ASBH) is a
professional society of more than 1 500 individuals, organisations,
and institutions interested in bioethics and humanities. The
website serves as a source of information for anyone interested
in bioethics.

Bioethics Resources on the Web,
http ://www.nih.gov/sigs/bioethics/
This website contains a broad range of web links, providing
background information and various positions on issues in
bioethics.

Centre for Bioethics and Health Law, http ://www.uu-cbg.nl/
An independent academic institute at Utrecht University offering
advice and research. It organises courses and provides training
sessions. Topics of particular interest for the Centre for
Bioethics and Health Law (CBG) are ethical and legal aspects of
biotechnology and genetic modification of humans, animals
and plants. Ethical questions in health care are a major field of
interest.

Clinical Trials, http ://clinicaltrials.gov
Provides regularly updated information about clinical research
in human volunteers. *clinicaltrials.gov* gives you information
about a trial's purpose and who may participate. Also responds
to lists of frequently asked questions concerning clinical trials.

Bioethics division of the Council of Europe,
http ://www.coe.int/T/E/legal_affairs/Legal_cooperation/bioethics/
The major bioethical issues examined from a European point of
view, and the essential legal texts.

European Union,
http ://europa.eu.int/comm/research/biosociety/bioethics/
bioethics_ethics_en.htm
The website contains a list of all national bioethics committees
for all EU member states.

Généthique, http://www.genethique.org/en.htm
The objective of the Généthique Forum is to be a centre for online discussion and information exchange for scientists, doctors and politicians in their quest for an objective reflection in various scientific domains.

German Reference Centre for Ethics in the Life Sciences, www.drze.de/links
An extensive list of bioethics links which enables easy access to a wide range of Internet sites in the area of ethics in the life sciences.

International Calendar for Bioethics Events, http://www2.umdnj.edu/ethicweb/upcome.htm
Regularly updated online calendar for bioethics forums, conferences and debates worldwide.

Johns Hopkins University Bioethics Institute, http://www.hopkinsmedicine.org/bioethics/
The Phoebe R. Berman Bioethics Institute of the Johns Hopkins University seeks to promote research in bioethics and encourage moral reflection among a broad range of scholars, professionals, students, and citizens. The institute serves the entire Johns Hopkins University and enables students and trainees to advance their understanding of bioethics in their personal and professional lives.

Midwest Bioethics Center, http://www.midbio.org/
An independent practical bioethics centre dedicated to raising and responding to ethical issues in health and health care.

National Consultative Bioethics Committee, http://www.ccne-ethique.org/
To give opinions on ethical problems raised by progress in the fields of biology, medicine and health, and to publish recommendations on the subject.

The United States National Library of Medicine, http://www.nlm.nih.gov/pubs/cbm/hum_exp.html
The NLM compiled a comprehensive bibliography, from 1989 to November 1998, entitled "Ethical issues in research involving human participants".

National Reference Center for Bioethics Literature,
www.georgetown.edu/research/nrcbl/nrc/
The National Reference Center for Bioethics Literature (NRCBL) is part of the Kennedy Institute of Ethics, Georgetown University. It is a specialised collection of books, journals, newspaper articles, legal materials, regulations, codes, government publications and other relevant documents concerned with issues in biomedical and professional ethics. The site includes resources on ethics and human genetics, including a searchable database of bibliographic references. The site also provides free access to Bioethicsline, a database of bioethical literature.

NIH Office of Extramural Research (OER), and the NIH Inter-Institute Bioethics Interest Group,
http://www.nih.gov/sigs/bioethics/
The OER provides information on policies and regulations, resources, guidance for clinical investigators, research resources, courses and tutorials on bioethical issues in human studies.

Nuffield Council on Bioethics,
http://www.nuffieldbioethics.org/home/
An independent body established to consider the ethical issues arising from developments in medicine and biology. The Council plays a major role in contributing to policy-making and stimulating debate in bioethics.

HHS Office of Human Research Protection,
http://ohrp.osophs.dhhs.gov/polasur.htm
The OHRP provides a guide and training materials on regulations and procedures governing research with human subjects; it includes a guidance document on financial relationships in clinical research.

The Stanford University Center for Biomedical Ethics,
http://scbe.stanford.edu/
The SCBE engages in interdisciplinary research on moral questions arising from the complex relationships among medicine, science and society. Its research is committed to exploring and promoting compassionate approaches to the practice of medicine.

Unesco International Bioethics Committee,
http ://portal.Unesco.org/shs/en/ev.php@URL_ID=1879&URL_
DO=DO_TOPIC&URL_SECTION=201.html
The IBC is a body of thirty-six independent experts that follows
progress in the life sciences and provides the only global forum
for in-depth bioethical reflection by exposing the issues at
stake.

University of Pennsylvania Bioethics Center,
http ://www.bioethics.upenn.edu/
The Center for Bioethics is a leader in bioethics research and its
deployment in the ethical practice of the life sciences and medicine. The center is a world-renowned educational and research
enterprise. Includes a useful link to Bioethics for Beginners.

University of Toronto Joint Centre for Bioethics,
http ://www.utoronto.ca/jcb/
The JCB is a partnership between the University of Toronto
and affiliated hospitals. The JCB studies important ethical,
health-related topics through research and clinical activities
and seeks to improve healthcare standards at both national
and international levels. It aims to provide leadership in
bioethics research, education, and clinical activities.

World Medical Association Ethics Unit,
http ://www.wma.net/e/ethicsunit/index.htm
This website includes information on the Declaration of
Helsinki and other ethical principles involved in health
research.

Appendix II – Additional protocol on biomedical research, Council of Europe

Additional Protocol to the Convention for the Protection of Human Rights and Dignity of the Human Being with regard to the Application of Biology and Medicine, on Biomedical Research

Preamble

The member States of the Council of Europe, the other States and the European Community signatories to this Additional Protocol to the Convention for the Protection of Human Rights and Dignity of the Human Being with regard to the Application of Biology and Medicine (hereinafter referred to as "the Convention"),

Considering that the aim of the Council of Europe is the achievement of greater unity between its members and that one of the methods by which this aim is pursued is the maintenance and further realisation of human rights and fundamental freedoms;

Considering that the aim of the Convention, as defined in Article 1, is to protect the dignity and identity of all human beings and guarantee everyone, without discrimination, respect for their integrity and other rights and fundamental freedoms with regard to the application of biology and medicine;

Considering that progress in medical and biological sciences, in particular advances obtained through biomedical research, contributes to saving lives and improving quality of life;

Conscious of the fact that the advancement of biomedical science and practice is dependent on knowledge and discovery which necessitates research on human beings;

Stressing that such research is often transdisciplinary and international;

Taking into account national and international professional standards in the field of biomedical research and the previous work of the Committee of Ministers and the Parliamentary Assembly of the Council of Europe in this field;

Convinced that biomedical research that is contrary to human dignity and human rights should never be carried out;

Stressing the paramount concern to be the protection of the human being participating in research;

Affirming that particular protection shall be given to human beings who may be vulnerable in the context of research;

Recognising that every person has a right to accept or refuse to undergo biomedical research and that no one should be forced to undergo such research;

Resolving to take such measures as are necessary to safeguard human dignity and the fundamental rights and freedoms of the individual with regard to biomedical research,

Have agreed as follows:

CHAPTER I

Object and scope

Article 1 – Object and purpose

Parties to this Protocol shall protect the dignity and identity of all human beings and guarantee everyone, without discrimination, respect for their integrity and other rights and fundamental freedoms with regard to any research involving interventions on human beings in the field of biomedicine.

Article 2 – Scope

1. This Protocol covers the full range of research activities in the health field involving interventions on human beings.

2. This Protocol does not apply to research on embryos *in vitro*. It does apply to research on foetuses and embryos *in vivo*.

3. For the purposes of this Protocol, the term "intervention" includes:

i. a physical intervention, and

ii. any other intervention in so far as it involves a risk to the psychological health of the person concerned.

CHAPTER II

General provisions

Article 3 – Primacy of the human being

The interests and welfare of the human being participating in research shall prevail over the sole interest of society or science.

Article 4 – General rule

Research shall be carried out freely, subject to the provisions of this Protocol and the other legal provisions ensuring the protection of the human being.

Article 5 – Absence of alternatives

Research on human beings may only be undertaken if there is no alternative of comparable effectiveness.

Article 6 – Risks and benefits

1. Research shall not involve risks and burdens to the human being disproportionate to its potential benefits.

2. In addition, where the research does not have the potential to produce results of direct benefit to the health of the research participant, such research may only be undertaken if the research entails no more than acceptable risk and acceptable burden for the research participant. This shall be without prejudice to the provision contained in Article 15 paragraph 2, subparagraph ii for the protection of persons not able to consent to research.

Article 7 – Approval

Research may only be undertaken if the research project has been approved by the competent body after independent examination of its scientific merit, including assessment of the importance of the aim of research, and multidisciplinary review of its ethical acceptability.

Article 8 – Scientific quality

Any research must be scientifically justified, meet generally accepted criteria of scientific quality and be carried out in accordance with relevant professional obligations and standards under the supervision of an appropriately qualified researcher.

CHAPTER III

Ethics committee

Article 9 – Independent examination by an ethics committee

1. Every research project shall be submitted for independent examination of its ethical acceptability to an ethics committee. Such projects shall be submitted to independent examination in each State in which any research activity is to take place.

2. The purpose of the multidisciplinary examination of the ethical acceptability of the research project shall be to protect the dignity, rights, safety and well-being of research participants. The assessment of the ethical acceptability shall draw on an appropriate range of expertise and experience adequately reflecting professional and lay views.

3. The ethics committee shall produce an opinion containing reasons for its conclusion.

Article 10 – Independence of the ethics committee

1. Parties to this Protocol shall take measures to assure the independence of the ethics committee. That body shall not be subject to undue external influences.

2. Members of the ethics committee shall declare all circumstances that might lead to a conflict of interest. Should such conflicts arise, those involved shall not participate in that review.

Article 11 – Information for the ethics committee

1. All information which is necessary for the ethical assessment of the research project shall be given in written form to the ethics committee.

2. In particular, information on items contained in the appendix to this Protocol shall be provided, in so far as it is relevant for the research project. The appendix may be amended by the Committee set up by Article 32 of the Convention by a two-thirds majority of the votes cast.

Article 12 – Undue influence

The ethics committee must be satisfied that no undue influence, including that of a financial nature, will be exerted on persons to participate in research. In this respect, particular attention must be given to vulnerable or dependent persons.

CHAPTER IV

Information and consent

Article 13 – Information for research participants

1. The persons being asked to participate in a research project shall be given adequate information in a comprehensible form. This information shall be documented.

2. The information shall cover the purpose, the overall plan and the possible risks and benefits of the research project, and include the opinion of the ethics committee. Before being asked to consent to participate in a research project, the persons concerned shall be specifically informed, according to the nature and purpose of the research:

i.   of the nature, extent and duration of the procedures involved, in particular, details of any burden imposed by the research project;

ii. of available preventive, diagnostic and therapeutic procedures;

iii. of the arrangements for responding to adverse events or the concerns of research participants;

iv. of arrangements to ensure respect for private life and ensure the confidentiality of personal data;

v.   of arrangements for access to information relevant to the participant arising from the research and to its overall results;

vi. of the arrangements for fair compensation in the case of damage;

vii. of any foreseen potential further uses, including commercial uses, of the research results, data or biological materials;

viii. of the source of funding of the research project.

3. In addition, the persons being asked to participate in a research project shall be informed of the rights and safeguards prescribed by law for their protection, and specifically of their right to refuse consent or to withdraw consent at any time without being subject to any form of discrimination, in particular regarding the right to medical care.

Article 14 – Consent

1. No research on a person may be carried out, subject to the provisions of both Chapter V and Article 19, without the informed, free, express, specific and documented consent of the person. Such consent may be freely withdrawn by the person at any phase of the research.

2. Refusal to give consent or the withdrawal of consent to participation in research shall not lead to any form of discrimination against the person concerned, in particular regarding the right to medical care.

3. Where the capacity of the person to give informed consent is in doubt, arrangements shall be in place to verify whether or not the person has such capacity.

CHAPTER V

Protection of persons not able to consent to research

Article 15 – Protection of persons not able to consent to research

1. Research on a person without the capacity to consent to research may be undertaken only if all the following specific conditions are met:

i.   the results of the research have the potential to produce real and direct benefit to his or her health;

ii. research of comparable effectiveness cannot be carried out on individuals capable of giving consent;

iii. the person undergoing research has been informed of his or her rights and the safeguards prescribed by law for his or her protection, unless this person is not in a state to receive the information;

iv. the necessary authorisation has been given specifically and in writing by the legal representative or an authority, person or body provided for by law, and after having received the information required by Article 16, taking into account the person's previously expressed wishes or objections. An adult not able to consent shall as far as possible take part in the authorisation procedure. The opinion of a minor shall be taken into consideration as an increasingly determining factor in proportion to age and degree of maturity;

v.   the person concerned does not object.

2. Exceptionally and under the protective conditions prescribed by law, where the research has not the potential to produce results of direct benefit to the health of the person concerned, such research may be authorised subject to the conditions laid down in paragraph 1, sub-paragraphs ii, iii, iv, and v above, and to the following additional conditions:

i. the research has the aim of contributing, through significant improvement in the scientific understanding of the individual's condition, disease or disorder, to the ultimate attainment of results capable of conferring benefit to the person concerned or to other persons in the same age category or afflicted with the same disease or disorder or having the same condition;

ii. the research entails only minimal risk and minimal burden for the individual concerned; and any consideration of additional potential benefits of the research shall not be used to justify an increased level of risk or burden.

3. Objection to participation, refusal to give authorisation or the withdrawal of authorisation to participate in research shall not lead to any form of discrimination against the person concerned, in particular regarding the right to medical care.

Article 16 – Information prior to authorisation

1. Those being asked to authorise participation of a person in a research project shall be given adequate information in a comprehensible form. This information shall be documented.

2. The information shall cover the purpose, the overall plan and the possible risks and benefits of the research project, and include the opinion of the ethics committee. They shall further be informed of the rights and safeguards prescribed by law for the protection of those not able to consent to research and specifically of the right to refuse or to withdraw authorisation at any time, without the person concerned being subject to any form of discrimination, in particular regarding the right to medical care. They shall be specifically informed according to the nature and purpose of the research of the items of information listed in Article 13.

3. The information shall also be provided to the individual concerned, unless this person is not in a state to receive the information.

Article 17 – Research with minimal risk and minimal burden

1. For the purposes of this Protocol it is deemed that the research bears a minimal risk if, having regard to the nature and scale of the intervention, it is to be expected that it will result, at the most, in a very slight and temporary negative impact on the health of the person concerned.

2. It is deemed that it bears a minimal burden if it is to be expected that the discomfort will be, at the most, temporary and very slight for the person concerned. In assessing the burden for an individual, a person enjoying the special confidence of the person concerned shall assess the burden where appropriate.

CHAPTER VI

Specific situations

Article 18 – Research during pregnancy or breastfeeding

1. Research on a pregnant woman which does not have the potential to produce results of direct benefit to her health, or to that of her embryo, foetus or child after birth, may only be undertaken if the following additional conditions are met:

i.   the research has the aim of contributing to the ultimate attainment of results capable of conferring benefit to other women in relation to reproduction or to other embryos, foetuses or children;

ii. research of comparable effectiveness cannot be carried out on women who are not pregnant;

iii. the research entails only minimal risk and minimal burden.

2. Where research is undertaken on a breastfeeding woman, particular care shall be taken to avoid any adverse impact on the health of the child.

Article 19 – Research on persons in emergency clinical situations

1. The law shall determine whether, and under which protective additional conditions, research in emergency situations may take place when:

i.   a person is not in a state to give consent, and

ii. because of the urgency of the situation, it is impossible to obtain in a sufficiently timely manner, authorisation from his or her representative or an authority or a person or body which would in the absence of an emergency situation be called upon to give authorisation.

2. The law shall include the following specific conditions:

i. research of comparable effectiveness cannot be carried out on persons in non-emergency situations;

ii. the research project may only be undertaken if it has been approved specifically for emergency situations by the competent body;

iii. any relevant previously expressed objections of the person known to the researcher shall be respected;

iv. where the research has not the potential to produce results of direct benefit to the health of the person concerned, it has the aim of contributing, through significant improvement in the scientific understanding of the individual's condition, disease or disorder, to the ultimate attainment of results capable of conferring benefit to the person concerned or to other persons in the same category or afflicted with the same disease or disorder or having the same condition, and entails only minimal risk and minimal burden.

3. Persons participating in the emergency research project or, if applicable, their representatives shall be provided with all the relevant information concerning their participation in the research project as soon as possible. Consent or authorisation for continued participation shall be requested as soon as reasonably possible.

Article 20 – Research on persons deprived of liberty

Where the law allows research on persons deprived of liberty, such persons may participate in a research project in which the results do not have the potential to produce direct benefit to their health only if the following additional conditions are met:

i. research of comparable effectiveness cannot be carried out without the participation of persons deprived of liberty;

ii. the research has the aim of contributing to the ultimate attainment of results capable of conferring benefit to persons deprived of liberty;

iii. the research entails only minimal risk and minimal burden.

CHAPTER VII

Safety and supervision

Article 21 – Minimisation of risk and burden

1. All reasonable measures shall be taken to ensure safety and to minimise risk and burden for the research participants.

2. Research may only be carried out under the supervision of a clinical professional who possesses the necessary qualifications and experience.

Article 22 – Assessment of health status

1. The researcher shall take all necessary steps to assess the state of health of human beings prior to their inclusion in research, to ensure that those at increased risk in relation to participation in a specific project be excluded.

2. Where research is undertaken on persons in the reproductive stage of their lives, particular consideration shall be given to the possible adverse impact on a current or future pregnancy and the health of an embryo, foetus or child.

Article 23 – Non-interference with necessary clinical interventions

1. Research shall not delay nor deprive participants of medically necessary preventive, diagnostic or therapeutic procedures.

2. In research associated with prevention, diagnosis or treatment, participants assigned to control groups shall be assured of proven methods of prevention, diagnosis or treatment.

3. The use of placebo is permissible where there are no methods of proven effectiveness, or where withdrawal or withholding of such methods does not present an unacceptable risk or burden.

Article 24 – New developments

1. Parties to this Protocol shall take measures to ensure that the research project is re-examined if this is justified in the light of scientific developments or events arising in the course of the research.

2. The purpose of the re-examination is to establish whether:

i. the research needs to be discontinued or if changes to the research project are necessary for the research to continue;

ii. research participants, or if applicable their representatives, need to be informed of the developments or events;

iii. additional consent or authorisation for participation is required.

3. Any new information relevant to their participation shall be conveyed to the research participants, or, if applicable, to their representatives, in a timely manner.

4. The competent body shall be informed of the reasons for any premature termination of a research project.

CHAPTER VIII

Confidentiality and right to information

Article 25 – Confidentiality

1. Any information of a personal nature collected during biomedical research shall be considered as confidential and treated according to the rules relating to the protection of private life.

2. The law shall protect against inappropriate disclosure of any other information related to a research project that has been submitted to an ethics committee in compliance with this Protocol.

Article 26 – Right to information

1. Research participants shall be entitled to know any information collected on their health in conformity with the provisions of Article 10 of the Convention.

2. Other personal information collected for a research project will be accessible to them in conformity with the law on the protection of individuals with regard to processing of personal data.

Article 27 – Duty of care

If research gives rise to information of relevance to the current or future health or quality of life of research participants, this information must be offered to them. That shall be done within a framework of health care or counselling. In communication of such information, due care must be taken in order to protect confidentiality and to respect any wish of a participant not to receive such information.

Article 28 – Availability of results

1. On completion of the research, a report or summary shall be submitted to the ethics committee or the competent body.

2. The conclusions of the research shall be made available to participants in reasonable time, on request.

3. The researcher shall take appropriate measures to make public the results of research in reasonable time.

CHAPTER IX

Research in States not parties to this Protocol

Article 29 – Research in States not parties to this Protocol

Sponsors or researchers within the jurisdiction of a Party to this Protocol that plan to undertake or direct a research project in a State not party to this Protocol shall ensure that, without prejudice to the provisions applicable in that State, the research project complies with the principles on which the provisions of this Protocol are based. Where necessary, the Party shall take appropriate measures to that end.

CHAPTER X

Infringement of the provisions of the Protocol

Article 30 – Infringement of the rights or principles

The Parties shall provide appropriate judicial protection to prevent or to put a stop to an unlawful infringement of the rights or principles set forth in this Protocol at short notice.

Article 31 – Compensation for damage

The person who has suffered damage as a result of participation in research shall be entitled to fair compensation according to the conditions and procedures prescribed by law.

Article 32 – Sanctions

Parties shall provide for appropriate sanctions to be applied in the event of infringement of the provisions contained in this Protocol.

CHAPTER XI

Relation between this Protocol and other provisions and re-examination of the Protocol

Article 33 – Relation between this Protocol and the Convention

As between the Parties, the provisions of Articles 1 to 32 of this Protocol shall be regarded as additional articles to the Convention, and all the provisions of the Convention shall apply accordingly.

Article 34 – Wider protection

None of the provisions of this Protocol shall be interpreted as limiting or otherwise affecting the possibility for a Party to grant research participants a wider measure of protection than is stipulated in this Protocol.

Article 35 – Re-examination of the Protocol

In order to monitor scientific developments, the present Protocol shall be examined within the Committee referred to in Article 32 of the Convention no later than five years from the entry into force of this Protocol and thereafter at such intervals as the Committee may determine.

CHAPTER XII

Final clauses

Article 36 – Signature and ratification

This Protocol shall be open for signature by Signatories to the Convention. It is subject to ratification, acceptance or approval.

A Signatory may not ratify, accept or approve this Protocol unless it has previously or simultaneously ratified, accepted or approved the Convention. Instruments of ratification, acceptance or approval shall be deposited with the Secretary General of the Council of Europe.

Article 37 – Entry into force

1. This Protocol shall enter into force on the first day of the month following the expiration of a period of three months after the date on which five States, including at least four member States of the Council of Europe, have expressed their consent to be bound by the Protocol in accordance with the provisions of Article 36.

2. In respect of any State which subsequently expresses its consent to be bound by it, the Protocol shall enter into force on the first day of the month following the expiration of a period of three months after the date of the deposit of the instrument of ratification, acceptance or approval.

Article 38 – Accession

1. After the entry into force of this Protocol, any State which has acceded to the Convention may also accede to this Protocol.

2. Accession shall be effected by the deposit with the Secretary General of the Council of Europe of an instrument of accession which shall take effect on the first day of the month following the expiration of a period of three months after the date of its deposit.

Article 39 – Denunciation

1. Any Party may at any time denounce this Protocol by means of a notification addressed to the Secretary General of the Council of Europe.

2. Such denunciation shall become effective on the first day of the month following the expiration of a period of three months after the date of receipt of such notification by the Secretary General.

Article 40 – Notifications

The Secretary General of the Council of Europe shall notify the member States of the Council of Europe, the European Community, any Signatory, any Party and any other State which has been invited to accede to the Protocol of:

a. any signature;

b. the deposit of any instrument of ratification, acceptance, approval or accession;

c. any date of entry into force of this Protocol in accordance with Articles 37 and 38;

d. any other act, notification or communication relating to this Protocol.

In witness whereof the undersigned, being duly authorised thereto, have signed this Protocol.

Done at Strasbourg, this ...., in English and in French, both texts being equally authentic, in a single copy which shall be deposited in the archives of the Council of Europe. The Secretary General of the Council of Europe shall transmit certified copies to each member State of the Council of Europe, to the non-member States which have participated in the elaboration of this Protocol, to any State invited to accede to the Convention and to the European Community.

Appendix to the Additional Protocol on Biomedical Research

Information to be given to the ethics committee

Information on the following items shall be provided to the ethics committee, in so far as it is relevant for the research project:

Description of the project

i. the name of the principal researcher, qualifications and experience of researchers and, where appropriate, the clinically responsible person, and funding arrangements;

ii.   the aim and justification for the research based on the latest state of scientific knowledge;

iii. methods and procedures envisaged, including statistical and other analytical techniques;

iv.   a comprehensive summary of the research project in lay language;

v.   a statement of previous and concurrent submissions of the research project for assessment or approval and the outcome of those submissions;

Participants, consent and information

vi.   justification for involving human beings in the research project;

vii. the criteria for inclusion or exclusion of the categories of persons for participation in the research project and how those persons are to be selected and recruited;

viii. reasons for the use or the absence of control groups;

ix.   a description of the nature and degree of foreseeable risks that may be incurred through participating in research;

x.   the nature, extent and duration of the interventions to be carried out on the research participants, and details of any burden imposed by the research project;

xi.   arrangements to monitor, evaluate and react to contingencies that may have consequences for the present or future health of research participants;

xii. the timing and details of information for those persons who would participate in the research project and the means proposed for provision of this information;

xiii. documentation intended to be used to seek consent or, in the case of persons not able to consent, authorisation for participation in the research project;

xiv. arrangements to ensure respect for the private life of those persons who would participate in research and ensure the confidentiality of personal data;

xv. arrangements foreseen for information which may be generated and be relevant to the present or future health of those persons who would participate in research and their family members;

Other information

xvi. details of all payments and rewards to be made in the context of the research project;

xvii. details of all circumstances that might lead to conflicts of interest that may affect the independent judgement of the researchers;

xviii. details of any foreseen potential further uses, including commercial uses, of the research results, data or biological materials;

xix. details of all other ethical issues, as perceived by the researcher;

xx. details of any insurance or indemnity to cover damage arising in the context of the research project.

The ethics committee may request additional information necessary for evaluation of the research project.

Appendix III – World Medical Association's Declaration of Helsinki [1]

Ethical Principles for Medical Research Involving Human Subjects

Adopted by the 18th WMA General Assembly, Helsinki, Finland, June 1964, and amended by the 29th WMA General Assembly, Tokyo, Japan, October 1975; 35th WMA General Assembly, Venice, Italy, October 1983; 41st WMA General Assembly, Hong Kong, September 1989; 48th WMA General Assembly, Somerset West, Republic of South Africa, October 1996; and the 52nd WMA General Assembly, Edinburgh, Scotland, October 2000. Note of Clarification on Paragraph 29 added by the WMA General Assembly, Washington 2002.

A – Introduction

1. The World Medical Association has developed the Declaration of Helsinki as a statement of ethical principles to provide guidance to physicians and other participants in medical research involving human subjects. Medical research involving human subjects includes research on identifiable human material or identifiable data.

2. It is the duty of the physician to promote and safeguard the health of the people. The physician's knowledge and conscience are dedicated to the fulfilment of this duty.

3. The Declaration of Geneva of the World Medical Association binds the physician with the words, "The health of my patient will be my first consideration," and the International Code of Medical Ethics declares that "A physician shall act only in the patient's interest when providing medical care which might have the effect of weakening the physical and mental condition of the patient".

4. Medical progress is based on research which ultimately must rest in part on experimentation involving human subjects.

5. In medical research on human subjects, considerations related to the well-being of the human subject should take precedence over the interests of science and society.

1.
The Declaration of Helsinki (Document 17.C) is an official policy document of the World Medical Association, the global representative body for physicians. It was first adopted in 1964 (Helsinki, Finland) and revised in 1975 (Tokyo, Japan), 1983 (Venice, Italy), 1989 (Hong Kong), 1996 (Somerset-West, South Africa) and 2000 (Edinburgh, Scotland). Note of clarification on Paragraph 29 added by the WMA General Assembly, Washington 2002.

6. The primary purpose of medical research involving human subjects is to improve prophylactic, diagnostic and therapeutic procedures and the understanding of the aetiology and pathogenesis of disease. Even the best proven prophylactic, diagnostic and therapeutic methods must continuously be challenged through research for their effectiveness, efficiency, accessibility and quality.

7. In current medical practice and in medical research, most prophylactic, diagnostic and therapeutic procedures involve risks and burdens.

8. Medical research is subject to ethical standards that promote respect for all human beings and protect their health and rights. Some research populations are vulnerable and need special protection. The particular needs of the economically and medically disadvantaged must be recognised. Special attention is also required for those who cannot give or refuse consent for themselves, for those who may be subject to giving consent under duress, for those who will not benefit personally from the research and for those for whom the research is combined with care.

9. Research Investigators should be aware of the ethical, legal and regulatory requirements for research on human subjects in their own countries as well as applicable international requirements. No national ethical, legal or regulatory requirement should be allowed to reduce or eliminate any of the protections for human subjects set forth in this Declaration.

B – BASIC PRINCIPLES FOR ALL MEDICAL RESEARCH

10. It is the duty of the physician in medical research to protect the life, health, privacy and dignity of the human subject.

11. Medical research involving human subjects must conform to generally accepted scientific principles, be based on a thorough knowledge of the scientific literature, other relevant sources of information, and on adequate laboratory and, where appropriate, animal experimentation.

12. Appropriate caution must be exercised in the conduct of research which may affect the environment, and the welfare of animals used for research must be respected.

13. The design and performance of each experimental procedure involving human subjects should be clearly formulated in an experimental protocol. This protocol should be submitted for consideration, comment, guidance – and, where appropriate, approval – to a specially appointed ethical review committee, which must be independent of the investigator, the sponsor or any other kind of undue influence. This independent committee should be in conformity with the laws and regulations of the country in which the research experiment is performed. The committee has the right to monitor ongoing trials. The researcher has the obligation to provide monitoring information to the committee, especially any serious adverse events. The researcher should also submit to the committee, for review, information regarding funding, sponsors, institutional affiliations, other potential conflicts of interest and incentives for subjects.

14. The research protocol should always contain a statement of the ethical considerations involved and should indicate that there is compliance with the principles enunciated in this Declaration.

15. Medical research involving human subjects should be conducted only by scientifically qualified persons and under the supervision of a clinically competent medical person. The responsibility for the human subject must always rest with a medically qualified person and never rest on the subject of the research, even though the subject has given consent.

16. Every medical research project involving human subjects should be preceded by careful assessment of predictable risks and burdens in comparison with foreseeable benefits to the subject or to others. This does not preclude the participation of healthy volunteers in medical research. The design of all studies should be publicly available.

17. Physicians should abstain from engaging in research projects involving human subjects unless they are confident that the risks involved have been adequately assessed and can be

satisfactorily managed. Physicians should cease any investigation if the risks are found to outweigh the potential benefits or if there is conclusive proof of positive and beneficial results.

18. Medical research involving human subjects should only be conducted if the importance of the objective outweighs the inherent risks and burdens to the subject. This is especially important when the human subjects are healthy volunteers.

19. Medical research is only justified if there is a reasonable likelihood that the populations in which the research is carried out stand to benefit from the results of the research.

20. The subjects must be volunteers and informed participants in the research project.

21. The right of research subjects to safeguard their integrity must always be respected. Every precaution should be taken to respect the privacy of the subject, the confidentiality of the patient's information and to minimise the impact of the study on the subject's physical and mental integrity and on the personality of the subject.

22. In any research on human beings, each potential subject must be adequately informed of the aims, methods, sources of funding, any possible conflicts of interest, institutional affiliations of the researcher, the anticipated benefits and potential risks of the study and the discomfort it may entail. The subject should be informed of the right to abstain from participation in the study or to withdraw consent to participate at any time without reprisal. After ensuring that the subject has understood the information, the physician should then obtain the subjects freely-given informed consent, preferably in writing. If the consent cannot be obtained in writing, the non-written consent must be formally documented and witnessed.

23. When obtaining informed consent for the research project the physician should be particularly cautious if the subject is in a dependent relationship with the physician or may consent under duress. In that case the informed consent should be obtained by a well-informed physician who is not engaged in the investigation and who is completely independent of this relationship.

24. For a research subject who is legally incompetent, physically or mentally incapable of giving consent or is a legally incompetent minor, the investigator must obtain informed consent from the legally authorised representative in accordance with applicable law. These groups should not be included in research unless the research is necessary to promote the health of the population represented and this research cannot instead be performed on legally competent persons.

25. When a subject deemed legally incompetent, such as a minor child, is able to give assent to decisions about participation in research, the investigator must obtain that assent in addition to the consent of the legally authorised representative.

26. Research on individuals from whom it is not possible to obtain consent, including proxy or advance consent, should be done only if the physical/mental condition that prevents obtaining informed consent is a necessary characteristic of the research population. The specific reasons for involving research subjects with a condition that renders them unable to give informed consent should be stated in the experimental protocol for consideration and approval of the review committee. The protocol should state that consent to remain in the research should be obtained as soon as possible from the individual or a legally authorised surrogate.

27. Both authors and publishers have ethical obligations. In publication of the results of research, the investigators are obliged to preserve the accuracy of the results. Negative as well as positive results should be published or be otherwise publicly available. Sources of funding, institutional affiliations and any possible conflicts of interest should be declared in the publication. Reports of experimentation not in accordance with the principles laid down in this Declaration should not be accepted for publication.

C – ADDITIONAL PRINCIPLES FOR MEDICAL RESEARCH COMBINED WITH MEDICAL CARE

28. The physician may combine medical research with medical care, only to the extent that the research is justified by its

potential prophylactic, diagnostic or therapeutic value. When medical research is combined with medical care, additional standards apply to protect the patients who are research subjects.

29. The benefits, risks, burdens and effectiveness of a new method should be tested against those of the best current prophylactic, diagnostic and therapeutic methods. This does not exclude the use of placebo, or no treatment, in studies where no proven prophylactic, diagnostic or therapeutic method exists. (See footnote earlier)

30. At the conclusion of the study, every patient entered into the study should be assured of access to the best proven prophylactic, diagnostic and therapeutic methods identified by the study.

31. The physician should fully inform the patient which aspects of the care are related to the research. The refusal of a patient to participate in a study must never interfere with the patient–physician relationship.

32. In the treatment of a patient, where proven prophylactic, diagnostic and therapeutic methods do not exist or have been ineffective, the physician, with informed consent from the patient, must be free to use unproven or new prophylactic, diagnostic and therapeutic measures, if in the physician's judgement it offers hope of saving life, re-establishing health or alleviating suffering. Where possible, these measures should be made the object of research, designed to evaluate their safety and efficacy. In all cases, new information should be recorded and, where appropriate, published. The other relevant guidelines of this Declaration should be followed.

Footnote: Note of clarification on paragraph 29 of the WMA Declaration of Helsinki

The WMA hereby reaffirms its position that extreme care must be taken in making use of a placebo-controlled trial and that in general this methodology should only be used in the absence of existing proven therapy. However, a placebo-controlled trial may be ethically acceptable, even if proven therapy is available, under the following circumstances:

– Where for compelling and scientifically sound methodological reasons its use is necessary to determine the efficacy or safety of a prophylactic, diagnostic or therapeutic method; or

– Where a prophylactic, diagnostic or therapeutic method is being investigated for a minor condition and the patients who receive placebo will not be subject to any additional risk of serious or irreversible harm.

All other provisions of the Declaration of Helsinki must be adhered to, especially the need for appropriate ethical and scientific review.